L O G I

The Brain Behind the Brain

Learn

LOGIC & MATH

together

For ages 9+

Join Facebook	 **https://facebook.com/** **groups/logitica9**
Homeschoolers *Plants don't grow in a day, and neither does logic.*	The author offers a uniquely designed once a week assignment based course: **Logitica: Build the Logic.** Please contact the author for the 2 months FREE trial of the course. **https://facebook.com/** **groups/logitica9**

Table of Contents

Logitica: Learn Logica & Math Together

Chapter 1 - **Number Box** ...18

| 7 | 65 | 9 | | 11 | 65 | 3 | | 5 | ? | 9 |

Keywords: Arithmetic Operations, Binary Operators, Reasoning.

Chapter 2 - **Number Cross** ...36

```
      9              5             19
  4 (14) 2      10 (6) 8       5 (?) 1
      2              2             11
```

Keywords: Arithmetic Operations, Binary Operators, Reasoning.

Chapter 3 - **Marbles in a Box** ...66

19   **25**

○● = ?   ● = ?

Keywords: Linear Equations.

Chapter 4 - **Average Cell** ...81

6C

A B C A -8

B

Keywords: Linear Equations, Fractions, Arithmetic Mean.

Chapter 5 - **Wisgo Number Tile** ...95

25 41 57 ?

Keywords: Stimulating the left and right sides of the brain.

Chapter 6 - **Number Pyramid**

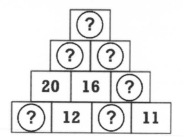

Keywords: Linear Equations, Pascal's Triangle.

Chapter 7 - **Advanced Number Pyramid**

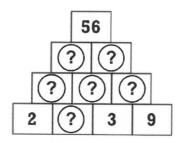

Keywords: Linear Equations, Pascal's Triangle.

Section-II Appendix

Section-III Answers

Publisher's note

Author

Neelabh Kumar is a thinker. Having memorized the first **1500 digits of Pi (π)** using sequential recollection, he is ranked among the top **150 on the Pi World Ranking List**. He is the creator behind Wisgo Logitica, which stimulates both sides of the brain. One of the Wisgo Logiticas Kumar created has a **patent filing in Hong Kong**. After earning a Masters Degree from one of the most prestigious universities in India (IIT), Kumar is now employed in Hong Kong at a large financial firm, while also creating and designing a new Logitica, with more to come.

Editors:

(1) Dilip Kumar Das, B.E. Civil Engineering, Retired executive from Visakhapatnam Steel Plant, India

(2) Suzanne M. Brierley, M.A., USA

(3) Reetu Das, B.Tech. (Computer Science), India.

Publisher's Cataloging-in-Publication data

Wisgo Limited, Hong Kong

Logitica: The Brain Behind the Brain

Learn LOGIC & MATH Together

First Edition 7.0

Dedication

I lovingly dedicate this book

to

my parents

Mr. Tarkeshwar Prasad & Late. Mridulata Verma

Thanks for your love and support.

Acknowledgment

I express my thanks to Mr. Dilip Kumar Das for editing and designing this book. A very special thanks to Mrs. Lipica Das for encouraging me to write this unique book.

I also thank my friends and family members for their suggestions and encouragement.

Thank you Reetu, my wife and best friend. Thanks for editing and presenting the book in a nice and easy to read format.

Three important concepts in Logitica

Logitica: The Brain Behind the Brain

I. Build true knowledge

Continuous learning, along with building logic iteratively, is **true knowledge.**

II. Don't memorize

Build the logic, not the memory.

III. Do memorize

Memorize the challenges posed by the problem, not its solution.

Preface

LOGITICA: The Brain Behind the Brain
Learn LOGIC & MATH together

> *Plants don't grow in a day, and neither does logic.*
>
> *Neelabh*

I. About this book and Logitica

In this book, we are going to introduce a new concept called *LOGITICA*. The focus of *Logitica* is to improve *critical thinking and problem-solving skills* by working through the challenges of solving a variety of problems. Each *Logitica* is carefully designed to stimulate iterative, analytical and logical thinking process.

LOGITICA can be considered as, "**The Brain Behind the Brain.**" Just like our brain tells us how to do things, *Logitica* is designed to train the brain how to think. The purpose of *Logitica* is described below in a few lines:

Purpose of Logitica: Logitica is a new concept, and new types of Logiticas are continually being created and designed by the author. However, this book is a good starting point for building logic in a unique way. Learning mathematics and logic together is an excellent way to build a foundation to develop one's cognitive abilities. When you combine this with problem-solving skills, you have the ability to rationally reason things out when presented with a problem and determine how to go forward in reaching the best possible solution. If you are reading this book, your never-ending journey towards intelligence is well underway. This journey, taken many times by creators and inventors in the past while solving unknown problems, is now accessible to an inquisitive and passionate mind through the innovation of **Logitica: The Brain Behind the Brain.**

II. Intended audience

This unique book is designed for 9+ year olds. The book is written with simple mathematical concepts, but the concepts may be slightly advanced for some kids. Hence, they might need parental or tutorial guidance while reading this book.

Now that the developed world has entered the *Digital Age*, it is important that we all adapt to the advancements being made in technology, which often require logical and analytical thinking. An excellent way to develop these skills is by learning the concepts taught in *Logitica* and putting those concepts to use while solving the problems in the book.

"LOGITICA: *Learn Logic & Math Together*" is the perfect textbook for private tutors and tutoring centers to use in helping their students develop analytical and logical thinking. These are the precise skills they will need to come out on top academically and succeed later on in their professional careers.

III. What is Logitica needed?

All parents want to give their children the best start in life and this means a good early education. Studies have shown that kids can learn complex concepts by the time they've mastered reading, which is around the ages of 8 or 9. Why not start now to promote the development of their cognitive abilities through the concepts taught in LOGITICA? When they learn logic and mathematics together it can build the foundation for what lies ahead in the 21st Century, a world where jobs in *Science, Technology, Engineering and Mathematics* (**STEM**) are in high demand and will continue to be.

Logitica specifically teaches how to approach different types of *mathematical problems* in a *logical manner* and presents the concepts in an engaging, fun and unique way. This helps to build logic in an iterative manner.

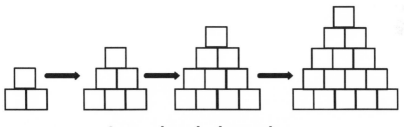

Learning is iterative

IV. Structure of each chapter

Step-1 Challenge: Each chapter starts with a *"Challenge Section"* containing the interesting problems that are designed to challenge your brain. In this section, we also provide the definition and objective of the problems.

Step-2 Strategy: To overcome the challenges of solving the problems, we provide you with the *"Strategy Section"* that contains *Analysis* and *Steps of Solving* the problems. Once we have acquired sufficient knowledge on the topic, we solve the problems individually by following the steps explained earlier.

Step-3 Answer: The answers to the problems in the chapter are summarized in the *Answer Section*.

3-steps mentioned above are described in the diagram below:

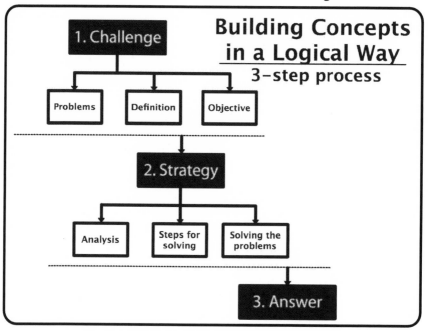

Figure-Build concepts in a logical way

Since analysis is the key to solving any problem, we perform a detailed analysis of the topic under discussion to understand the underlying principles required for solving the given problem.

Continuing on, sometimes it is possible to have more than one answer to a given problem. In some cases, only one answer may be provided. The reader is encouraged to find alternative answers, when possible. Each chapter ends with an Exercise Section, which contains a list of problems. The *Exercise Section* includes the following two levels of problems:

✔ **Level-I:** These are the more straightforward problems for you to solve and are useful in building a good basic understanding of the concepts discussed in the chapter.

✔ **Level-II:** The problems in this category are slightly more challenging than those in **Level-I**. Readers will learn more about the relevant *Logitica* by solving problems on this level.

The Appendix contains the derivation of some of the mathematical concepts used in The various chapters. We will refer to the relevant section if and when necessary.

V. Notion used in this book

We used the following symbols/notations in this book:

(i) Multiplication (✕): We will use this symbol for multiplication.

(ii) Division: We will use the standard symbol like $\dfrac{2}{3}$ to indicate division. .

(iii) ♟ Note: We will use this symbol to describe *important notes* on the relevant topic.

(iv) ♛ Definition: We will use this symbol to define the *important concepts* in the relevant chapter.

(v) ♞ Rule of thumb: We will use this symbol to describe some *important rules or clues* to help solve the relevant Logitica.

VI. How to read this book

The best way to read this book is to go through each chapter sequentially as arranged in the book. In general, you should spend two or more weeks on each *Logitica*. The reader is recommended to read the *Challenge and Strategy Sections* to understand the problem given in the chapter. Subsequently, the reader should read the *Analysis Section* to learn to solve the corresponding problem in the chapter. Once a good level of understanding is achieved, readers should read how to solve the problems given in the chapter. Since learning is an iterative process and it takes time to build the logic. So, you should not speed read through all the chapters at once, rather you should take your time to study and analyze the material in each chapter. If you are not able to fully understand a chapter, it is okay if you want to skip the chapter and come back later to learn it.

Referring to a Logitica: In simple terms, a specific *Logitica* generally refers to a category of problems that may span over multiple chapters. However, in this book, we have one chapter dedicated to one Logitica. Therefore, we will use the term "*Logitica*" or "*problem*" interchangeably whenever appropriate to mean the same thing.

Section-I

LOGITICA™
The Brain Behind the Brain

Learn

LOGIC & MATH

together

For ages 9+

Chapter 1

Number Box

Keywords: Arithmetic Operations, Binary Operators, Reasoning.

1.1 Challenge

Solve the following *Number Box* problems:

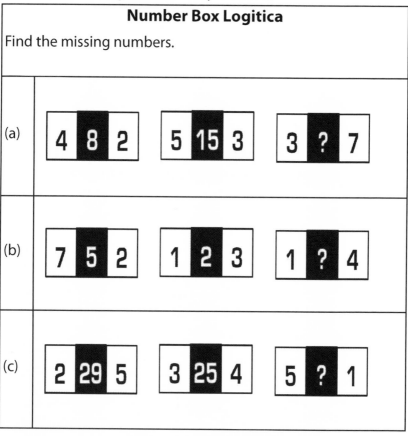

Number Box Logitica		
Find the missing numbers.		

(a) 4 8 2   5 15 3   3 ? 7

(b) 7 5 2   1 2 3   1 ? 4

(c) 2 29 5   3 25 4   5 ? 1

Figure 1.1: Problems

In this chapter, we will discuss various *arithmetic operations* and learn how to use them to solve problems involving three *Number Boxes*. Let us first define what we mean by a *Number Box*.

♛ Definition - Number Box

A **Number Box** is a box with three compartments: *left, middle, and right*. Each of these compartments contains a number as shown in figure-1.2. The middle number of each box is calculated by using *the numbers in the left* and *right compartments* of the box.

For example, as shown in figure-1.2, the middle number can be obtained by multiplying the number on the left by the number on the right: $4 \times 2 = 8$.

Objective: A *Number Box problem* contains three boxes, and all the numbers in the first two boxes have been provided. As a rule, the logic to determine the middle number must be the same for all of the boxes. Using this information, we need to find the missing number in the third box.

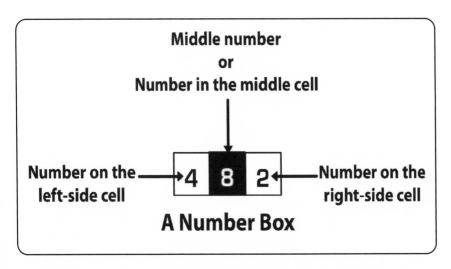

Figure 1.2: A box with 3 number compartments

Note:

(i) For brevity, throughout this chapter, we may refer to a *Number Box* as a *"box."*

(ii) We will refer to the three numbers in the box as *"number on the left," "middle number,"* and *"number on the right"* respectively.

1.2 Strategy

In this section, we will give an overview of a *Number Box* problem followed by a thorough analysis of the topic. We will later outline steps on how to solve a problem involving three *Number Boxes*.

1.2.1 Overview

As mentioned before, the logic to determine the middle number must be the same for all of the boxes, we need to figure out how the middle number is calculated for the first two boxes and apply the same logic to determine the missing number in the third box.

♖ **Note: Logic or pattern in a Number Box**

As per the definition above, the middle number is obtained by applying a combination of arithmetic operations on the two numbers on the left and right sides of the box. So when we say *"find the missing number,"* it means that we are supposed to find the appropriate combinations of arithmetic operations, which when applied to the *numbers on the left and right*, will yield the middle number.

For example, as shown in the figure on the right, the middle number can be obtained by multiplying the number on the left by the number on the right: 4 × 2 = 8. So, when we ask you to

| 4 | 8 | 2 |

A Number Box

determine the missing number in this example, you need to find the fact that the middle number is obtained by multiplying the number on the left by the number on the right. Once you discover the logic in one box, verify the logic in the other box, and then use the same logic to find the missing number in the third box.

♞ Assumptions: Ignore sequential dependency

Please refer to *Appendix-A* for further detail on *sequential dependency*, which is a slightly advanced topic. However, in this book, we will ignore sequential dependency when solving *Number Box* problems.

1.2.2 Analysis

Since the number on the left and the number on the right are used to determine the middle number, we need to use a certain category of binary operators for solving a *Number Box problem*. Let us first define what we mean by binary operators.

♛ Definition - Binary operators

Binary operators are those operators, which require two operands. The well-known binary operators are addition, subtraction, multiplication, and division. There are also some other types of binary operators, such as *logical AND (&&)*, *logical OR (||)*, etc. However, in this book, we will only refer to the types of binary operators that have to do with arithmetic operations.

Let us discuss some of the arithmetic operations that are useful for solving a *Number Box* problem.

(i) Addition: Add the two numbers indicated on the left and right sides together to determine what the middle number is. For example, as shown in the

A Number Box

figure on the right, the middle number is equal to 3 + 4 = 7.

(ii) Subtraction: Subtract one number from another to determine the middle number. For example, in the figure on the right, the middle number is obtained by

A Number Box

subtracting the number on the right from the number on the left, i.e., the middle number is 5 − 3= 2.

Since subtraction is *non-commutative*, the result will change if we reverse the order of numbers in a subtraction problem. Therefore, depending on the problem, the number at the center may also be obtained by subtracting number on the left from the number on the right. Similarly, the number at the center may be obtained by subtracting the smaller (or larger) from the larger (or smaller) number.

(iii) Multiplication: Multiply the two numbers on the left and right sides together to calculate the middle number. For example, as shown in the figure on the

A Number Box

right, the middle number is equal to 2 × 5 = 10.

(iv) Division: Divide one number by another. For example, as shown in the figure on the right, we can calculate the middle number by dividing the number

A Number Box

on the right by the number on the left, i.e., the middle number = $\frac{20}{5} = 4$.

We know that division is *non-commutative*, hence the result will change if we reverse the order of numbers in a division problem. Therefore, depending on the problem, the number at the center may also be obtained by dividing the number on the left by the number on the right. Similarly, the number at the center may be obtained by dividing the smaller (or larger) by the larger (or smaller) number.

1.2.1.1 Some more arithmetic operations

There are also other kinds of arithmetic operations used in solving a *Number Box* problem. We are going to list some of them in this chapter. It is worth noting that the list of the arithmetic operations given here is not exhaustive and a given problem may use a different combination of arithmetic operations. One can progressively learn to find such combinations by solving and practicing these types of problems.

(i) Sum of squares: The middle number could be the result of the number on the left squared plus the number on the right squared. As shown in figure here, the middle number is equal to $4^2 + 2^2 = 16 + 4$ = 20. In addition to this, there may be many other possible combinations, such

| 4 | 20 | 2 |

A Number Box

as, finding the difference between the squares or multiplying the squares together or calculating the average of squares, etc.

(ii) Sum of cubes: The middle number could be the result of the number on the left cubed plus the number on the right cubed. As shown in the figure here, the middle number is equal to $3^3 + 2^3 = 27 + 8 = 35$. In addition to this, there may be many other combinations possible, e.g., finding the

| 3 | 35 | 2 |

A Number Box

difference between cubes or multiplying the cubes together or calculating the average of cubes, etc.

(iii) Arithmetic Mean: An *arithmetic mean* is the average of a list of numbers. It is defined as the sum of the numbers divided by the count of these numbers. Here are the two examples on how to calculate the *arithmetic mean*:

(i) The *arithmetic mean* of 7 and 8 = $\dfrac{7+8}{2} = \dfrac{15}{2}$

(ii) The *arithmetic mean* of 1, 2, and 3 = $\dfrac{1+2+3}{3} = \dfrac{6}{3} = 2$

In a box, if the middle number is to be determined by averaging, it will be the average of the numbers on the left and right sides. As shown in the figure on the right, the middle number is $\dfrac{9+35}{2} = \dfrac{44}{2} = 22$.

$$\boxed{9 \;\; \blacksquare22\blacksquare \;\; 35}$$

A Number Box

1.2.3 Solving Steps

As we have already mentioned earlier, the middle number is obtained by applying some combinations of arithmetic operations on the two numbers on the left and right sides of the box. In the discussion below, we will use the term "logic" to refer to such combinations of arithmetic operations. We will follow the steps as described below to solve problems from this chapter:

✔ Review the problem by analyzing the first two boxes. Choose one of the first two boxes, and apply various arithmetic operations to the numbers on the left and right sides of the box. Find the appropriate combination of arithmetic operations, which would yield the number that equals the middle number of the chosen box.

✔ Now apply the same logic on the other box to confirm if the selected combination of arithmetic operations would yield the number that equals the middle number of the box.

✔ Once we have a logic that is valid for the first two boxes, apply the same logic to find the missing in the third box.

🐴 Rule of thumb - Start with simple arithmetic operations

In general, you should try using the simplest logic first, and if that does not work, try using another logic. The simplest logic that works is preferred. Here the simple logic is one that uses simple arithmetic operations.

1.2.4 Solving the Problems

After discussing the basics of a *Number Box problem*, let us look at the first problem.

Problem (a)

Figure 1.3: Problem (a)

We should attempt these types of problems by trying different arithmetic operations to see which one works:

• **Addition:**

The first box: 4 + 2 = 6. This result is not equal to the middle number 8.

• **Subtraction:**

The first box:

4 − 2 = 2. This result does not equal the middle number 8.

2 − 4 = −4. This result does not equal the middle

number 8.

- **Multiplication:**

    The first box: $4 \times 2 = 8$. This result equals the middle number 8.

    The second box: $5 \times 3 = 15$. This result equals the middle number 15.

    Since this logic works for the first two boxes, the missing number in the third box will be calculated by applying the same logic.

The third box: The missing number is $3 \times 7 = 21$

Now that we have found the answer to problem (a), which we will summarize later in the *Answer Section*. Let us discuss the next problem.

Problem (b)

Figure 1.4: Problem (b)

Let us first try a few arithmetic operations to see which one solves this problem:

- **Addition:**

    The first box: $7 + 2 = 9$. This result does not equal the middle number 2.

- **Subtraction:** Here we first try subtracting the smaller number from the bigger number:

    The first box: $7 - 2 = 5$. This result equals the middle number 5.

    The second box: $3 - 1 = 2$. This result equals the middle

number 2.

Since this logic works for the first two boxes, the missing number in the third box will be calculated by applying the same logic.

The third box: The missing number is 4 − 1 = 3

We will summarize the answer to problem (b) later in the *Answer Section*.

♖ Note:

(i) You cannot use 1 − 4= −3 in this case, because the middle numbers for the first two boxes are calculated by subtracting the smaller number from the bigger number.

(ii) As seen before, the middle number of the second box in problem (a) is the difference between two numbers: 3 − 1 = 2. However, did you notice that we can find the same middle number by averaging the numbers on the left and right sides of the box? Hence, the middle number can also be calculated as: $\dfrac{1+3}{2} = \dfrac{4}{2} = 2$. However, as mentioned before, whatever logic you select, it must be valid for the first two boxes before it can be used in finding the missing number of the third box. Since this logic is not valid for the first box, we can not use the average of numbers to find the missing number.

Second box from problem(b)

Problem (c)

| 2 | 29 | 5 | | 3 | 25 | 4 | | 5 | ? | 1 |

Figure 1.5: Problem (c)

Let us first try a few arithmetic operations to see which one solves this problem:

- **Addition:**

  The first box: $2 + 5 = 7$. This result does not equal the middle number 29.

- **Subtraction:**

  The first box:

  $2 - 5 = -3$. This result does not equal the middle number 29.

  $5 - 2 = 3$. This result does not equal the middle number 29.

- **Multiplication:**

  The first box: $2 \times 5 = 10$. This result does not equal the middle number 29.

After trying a few more combinations of arithmetic operations, we can find the answer by adding the squares of the numbers as shown below:

- **Sum of squares:**

  The first box: $2^2 + 5^2 = 4 + 25 = 29$. This result equals the middle number 29.

  The second box: $3^2 + 4^2 = 9 + 16 = 25$. This result equals the middle number 25.

  Hence we can calculate the missing number for the third box as:

The third box: The missing number is $5^2 + 1^2 = 25 + 1 = 26$.

Now that we have found the answers to all the questions, we will show them in the next section.

1.3 Answer

The answers to the problems are shown below:

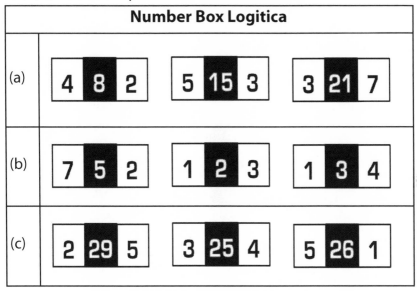

Number Box Logitica

(a) 4 8 2   5 15 3   3 21 7

(b) 7 5 2   1 2 3   1 3 4

(c) 2 29 5   3 25 4   5 26 1

Figure 1.6: Answers

1.4 Summary

Since we need only two numbers to find the middle number, a *Number Box problem* is easy to understand. The only complexity lies in finding which logic to use to determine the missing number. We will discuss another type of problem called the "*Number Cross*" in **Chapter-2** (p. 36), where we use four numbers to find the missing number.

We started this chapter with three problems and discussed how to determine the missing number using some well-known arithmetic operations. In each of the problems, the logic used to determine the middle number was different. However, the steps followed to find the logic were similar. We started with simple logic and continued applying different types of logic to the problem until we found one that was valid for the first two boxes. We then used that same logic to find the middle number in the third box. However, you may find there are more possible combinations of these operations when you are solving a *Number Box* problem.

A Number Box

In other words, to determine the middle number, we might need to apply various arithmetic operations on the two numbers on both sides of the middle number like $\dfrac{a^2+b^2}{2}$, $a^2 + b^3$, $a^2 - b^2$, $b(a - b)$, etc. In some cases, there are multiple answers possible, but as mentioned previously in this chapter, the logic that uses the simplest arithmetic and is consistent with all the boxes is preferred.

It is recommended that readers work on solving problems laid out in the *Exercise Section* to gain a better understanding of *Number Box problems*.

1.5 Exercise

Find the missing numbers in the problems below:

Level-I

(i)	12 **15** 3	-5 **-2** 3	9 **?** -10
(ii)	4 **20** 5	7 **91** 13	5 **?** 9
(iii)	5 **2** 3	7 **-4** 11	9 **?** 23
(iv)	5 **8** 11	19 **7** -5	11 **?** -7
(v)	5 **45** 9	9 **27** 3	4 **?** 7
(vi)	7 **4** 11	9 **5** 4	4 **?** 15
(vii)	5 **35** 2	3 **24** 5	7 **?** 1
(viii)	9 **20** 11	3 **12** 9	15 **?** 7
(ix)	9 **4** 5	7 **-4** 11	5 **?** 13
(x)	-5 **41** 4	2 **13** 3	4 **?** 9

(xi)	11	1	12	-15	24	9	-2	?	-4
(xii)	3	10	2	2	99	9	5	?	10
(xiii)	7	63	2	5	40	3	12	?	1
(xiv)	9	12	3	5	17	12	11	?	8
(xv)	5	50	5	3	21	4	3	?	5
(xvi)	4	4	5	5	35	12	3	?	5
(xvii)	5	-6	-3	-5	24	-3	9	?	-12
(xviii)	2	15	3	2	48	6	5	?	4
(xix)	2	24	4	7	-10	-2	1	?	3
(xx)	7	10	13	15	16	17	5	?	11
(xxi)	-7	-9	-11	3	4	5	-5	?	3

(xxii)	9 **7** 5	4 **6.5** 9	5 **?** 13
(xxiii)	7 **28** 11	8 **32** 4	-5 **?** -2
(xxiv)	3 **34** 5	4 **41** 5	3 **?** 9
(xxv)	1 **4** 4	5 **40** 8	3 **?** 5

Level-II

(i)	12 **6** 2	3 **4** 12	12 **?** 4
(ii)	1 **10** 3	3 **34** 5	5 **?** 4
(iii)	-5 **17** 3	-2 **34** -8	9 **?** -1
(iv)	2 **12** 8	-6 **-48** 2	5 **?** -3
(v)	-3 **19** -2	4 **56** 2	-4 **?** 2
(vi)	2 **-19** -3	2 **72** 4	6 **?** 7

(vii)	12 **2** 6	16 **4** 4	5 **?** -10
(viii)	1 **9** -2	3 **-2** 5	4 **?** -5
(ix)	5 **34** 3	3 **25** 4	1 **?** -4
(x)	1 **28** 3	3 **91** 4	2 **?** 5
(xi)	-2 **31** 3	-3 **-11** -4	-2 **?** -4
(xii)	3 **26** 1	-1 **63** -4	3 **?** -2
(xiii)	5 -5 -1	3 **4** 12	3 **?** -12
(xiv)	5 **61** -4	2 **35** 3	3 **?** 4
(xv)	2 **4** 8	16 **-2** -8	3 **?** 18
(xvi)	1 **26** 5	5 **52** 3	1 **?** 3
(xvii)	2 **6** 5	7 **28** 3	1 **?** 9

(xviii)	-5 **17** 3	6 **26** 4	5 **?** 7
(xix)	6 **50** 8	-9 **45** 3	2 **?** 6
(xx)	5 **36** 9	5 **6** 3	7 **?** 5
(xxi)	1 **-20** -4	2 **15** 5	-1 **?** 3
(xxii)	2 **33** 5	5 **89** 4	3 **?** 2
(xxiii)	3 **5** 1	5 **53** 9	9 **?** 3
(xxiv)	3 **31** 2	4 **89** -5	4 **?** 3
(xxv)	5 **-2** 3	-2 **17** 3	2 **?** -2

35

Chapter 2

Number Cross

Keywords: Arithmetic Operations, Binary Operators, Reasoning.

2.1 Challenge

Solve the following *Number Cross* problems:

Number Cross Logitica		
Find the missing numbers.		

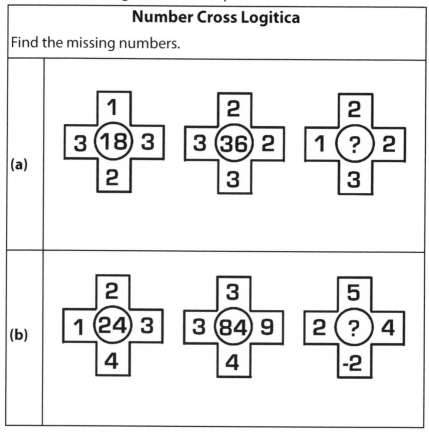

(a)

(b)

Figure 2.1: Problems

In this chapter, we will learn how to solve problems involving *Number Crosses*. Let us first define what we mean by a *Number Cross*.

♛ Definition - Number Cross

A **Number Cross** is a structure consisting of five numbers that are arranged in a cross-like shape as shown in the figure on the right. The number in the center of a cross is determined by applying some combinations of arithmetic operations on the four surrounding numbers.

A Number Cross

For example, as shown in the figure above, the number in the center is obtained by adding the four surrounding numbers: $1 + 3 + 5 + 4 = 13$.

Objective: The problems from this chapter consist of three crosses, and all the numbers in the first two crosses have been provided. Using this information, we need to find the missing number in the third cross. As a rule, the logic to determine the number in the center must be the same for all of the crosses.

♜ **Note**: For brevity, throughout this chapter, we may refer to a *Number Cross* as a *"cross."*

2.2 Strategy

In this section, we will first give an overview of a *Number Cross* followed by a detailed analysis of the topic. We will later outline steps describing how to solve problems involving *Number Crosses*.

2.2.1 Overview

As defined above, the number in the center is obtained by applying some combinations of arithmetic operations on the four surrounding numbers. So when we are asked to "*find the missing number*," it means that we are supposed to find the appropriate combinations of arithmetic operations, which when applied on the four surrounding numbers, will result in the same number that is in the center of the cross. For example, in the figure on the right, the number in the center is obtained by adding the four surrounding numbers: $1 + 3 + 5 + 4 = 13$. In this case,

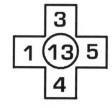

A Number Cross

"*determining the missing numbers*" requires you to find the fact that the number in the center is the "*sum of all four surrounding numbers*," and use the same logic to determine the missing number in the third cross. Since the logic used to determine the number in the center must be the same in all the three crosses, we need to validate the logic in the first two crosses before we can use that same logic in the third cross to find the answer.

2.2.2 Analysis

To solve problems involving *Number Crosses*, we need to use some of the concepts that we have already discussed while solving *Number Box* problems. So, if you have not read this yet, it is the time to go back and read about it in *Chapter-1* (p. 18). We have already discussed addition, subtraction, multiplication, division, and arithmetic mean, etc., in the context of solving *Number Box* problems. Now we are going to explore the same in the context of *Number Cross* problems.

To find the number in the center, we need to use some

arithmetic operations. As we know, *binary operators* require two operands, but a cross has four surrounding numbers. So we need to find a way to yield two numbers out of the four numbers seen in the cross. For this, we need to create two pairs of numbers out of the four surrounding numbers. To understand how numbers can be selected for pairings, we first need to understand the structure of a cross. A cross consists of two numbers in a horizontal row intersecting with two numbers in a vertical column as shown in figure-2.2.

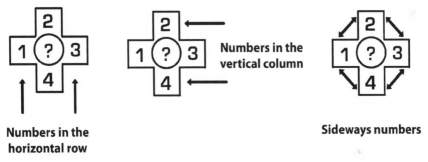

Figure 2.2: Structure of a Number Cross

We can create pairs in three different ways in a cross as listed below:

(a) Horizontal pair: The two numbers in the horizontal row are paired up. As shown in figure-2.2, the pair from the horizontal row is (1, 3).

(b) Vertical pair: The two numbers in the vertical column are paired up. As shown in figure-2.2, the pair from the vertical column is (2, 4).

(c) Sideways pair: Pairings of adjacent side numbers together are also possible. For example, in figure-2.2, the pairs made from the *sideways numbers* are (1, 2), (2, 3), (3, 4), and (4, 1).

2.2.1.1 An example of pairings in a cross

Let us follow the steps below to see how pairing up numbers can be used to find the number in the center of a cross:

✔ **Step 1**: Create two pairs of numbers using the surrounding numbers in the cross.

✔ **Step 2**: Apply various combinations of arithmetic operations to each pair, which will give us two new numbers.

✔ **Step 3**: We can now calculate the number in the center by using the two newly generated numbers from the previous step.

We will use the cross shown on the right to explain the steps discussed above and to demonstrate how we obtained 50 at the center:

A Number Cross

✔ **Step 1**: We create two pairs of numbers from the horizontal row and vertical column:

> Horizontal row: (2, 5)
>
> Vertical column: (3, 2)

✔ **Step 2**: We apply some combinations of arithmetic operations to the numbers in each pair to generate two new numbers (10, 5) as shown below:

> Horizontal row: $2 \times 5 = 10$
>
> Vertical column: $3 + 2 = 5$

✔ **Step 3**: We can now multiply the two new numbers to generate the number in the center, i.e., the number in the center is $10 \times 5 = 50$.

The above example demonstrates how a single number can be obtained by using the four surrounding numbers in the cross.

♞ Assumptions:

(i) Ignore sideways pairings in a cross

When solving *Number Cross problems*, we are going to ignore *sideways pairings*. This way we can focus on learning logic to solve the problem without worrying about the combinations of pairings possible in a cross. Aside from this, we can probably create a problem of similar complexity by just pairing the numbers from the horizontal row and vertical column. Readers should be aware that some books or certain tests might require the use of sideways pairings to find the solution, so prepare accordingly as per the *guidelines* and *norms* of those books or tests.

However, to ensure that we focus on learning the logic needed to solve the problem, we will exclude sideways pairings in this book.

(ii) Ignore sequential dependency

Please refer to **Appendix-B** for further detail on *sequential dependency*, which is a slightly advanced topic. However, in this book, we will ignore sequential dependency when solving *Number Cross* problems.

2.2.2.2 Arithmetic operations

Let us discuss some of the arithmetic operations that are useful in solving *Number Cross problems*:

(i) Addition: We can calculate the number in the center by adding up all the four surrounding numbers. As shown in the figure, the number in the center is equal to $2 + 3 + 5 + 7 = 17$. Besides this, other problems might use a variety of combinations, some of which

A Number Cross

are described below:

✔ Calculate the sum of two numbers from the horizontal row, which we can call A.

✔ Calculate the sum of two numbers from the vertical column, which we can call B.

✔ Once we have A and B, we can choose the appropriate method described in section **2.2.3.3 Calculate the number in the center using A and B** (p. 48) to find the number in the center. For example, we can use the four surrounding numbers in the figure above, and:

- Add numbers from the horizontal row: $A = 2 + 5 = 7$
- Add numbers from the vertical column: $B = 3 + 7 = 10$
- So, the number in the center can be obtained by using combinations like AB, $A - B$, $B - A$, $A^2 + B^2$,..., and so on. Here are a few examples for calculating the number in the center:

    - $B + A = 10 + 7 = 17$
    - $A - B = 7 - 10 = -3$
    - $B - A = 10 - 7 = 3$
    - $A^2 + B^2 = 7^2 + 10^2 = 49 + 100 = 149$, and so on.

(ii) Multiplication: The number in the center can be calculated by multiplying all four of the surrounding numbers together. As shown in the figure here, the number in the center is $3 \times 2 \times 4 \times 4 = 96$. Besides this, there are more possible combinations as described below:

A Number Cross

✔ Multiply the two numbers from the horizontal row, which we can call A.

✔ Multiply the two numbers from the vertical column, which we can call B.

✔ Once we have A and B, we can choose the appropriate method described in section **2.2.3.3 Calculate the number in the center using A and B** (p. 48) to find the number in the center. For example, we can use the four surrounding numbers in the figure above, and:

- multiply the two numbers from the horizontal row:

  $A = 3 \times 4 = 12$

- multiply the two numbers from the vertical column:

  $B = 2 \times 4 = 8$

- So, the number in the center can be obtained by using combinations like AB, $A - B$, $B - A$, $A^2 + B^2$,..., and so on. Here are a few examples:

  - $A + B = 12 + 8 = 20$
  - $A - B = 12 - 8 = 4$
  - $B - A = 8 - 12 = -4$
  - $A^2 + B^2 = 12^2 + 8^2 = 144 + 64 = 208$, and so on.

(iii) Subtraction: This is explained below with an example:

✔ Let us call the difference between two numbers from the horizontal row as A.

✔ Let us call the difference between two numbers from the vertical column as B.

✔ The number in the center in the figure here is $A - B$ as shown below:

Horizontal row: $A = 4 - 2 = 2$

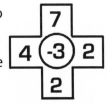

A Number Cross

Vertical column: $B = 7 - 2 = 5$

The number in the center $= A - B = 2 - 5 = -3$.

✔ In addition to the steps in above example, we can choose the appropriate method described in section **2.2.3.3 Calculate the number in the center using A and B** (p. 48) to find the number in the center. For example, we can use the four surrounding numbers in the figure above:

• Find the difference between numbers from the horizontal row:

$$A = (4 - 2) \text{ or } (2 - 4) = 2 \text{ or } -2$$

So, A has two possible values, $A = 2$ or -2.

• Find the difference between numbers from the vertical column:

$$B = (7 - 2) \text{ or } (2 - 7) = 5 \text{ or } -5$$

So, B also has two possible values, $B = 5$ or -5.

• Thus, the number in the center can be obtained by using combinations like AB, $A - B$, $B - A$, $A^2 + B^2$,..., and so on. Since subtraction is non-commutative, we can arrive at more values of the number in the center. Here are a few examples:

A + B : $2 + 5 = 7, -2 + 5 = 3, 2 - 5 = -3, -2 - 5 = -7$.

: $(7, 3, -3, -7)$

A − B : $2 - 5 = -3, -2 - 5 = -7, 2 + 5 = 7, -2 + 5 = 3$.

: $(-3, -7, 7, 3)$

AB : $2 \times 5 = 10, (-2) \times 5 = -10, 2 \times (-5) = -10, (-2) \times (-5) = 10$.

: $(10, -10)$

44

(iv) Division: This is explained below with an example:

✔ Divide one of the numbers in the horizontal row number by the other and call the result A.

✔ Divide one of the numbers in the vertical column number by the other and call the result B.

✔ The number in the center in the figure here is $\dfrac{A}{B}$ as shown below:

Horizontal row: $A = \dfrac{48}{4} = 12$

A Number Cross

Vertical column: $B = \dfrac{12}{4} = 3$

The number in the center $= \dfrac{A}{B} = \dfrac{12}{3} = 4$

✔ In addition to the steps mentioned above, once we have A and B, we can choose the appropriate method described in section **2.2.3.3 Calculate the number in the center using A and B** (p. 48) to find the number in the center. For example, using the four surrounding numbers in the figure above:

• Divide the numbers from the horizontal row with each other:

$$A = \frac{48}{4} = 12 \text{ or } A = \frac{4}{48} = \frac{1}{12}$$

So, A has two values possible, $A = (12, \dfrac{1}{12})$

• Divide the numbers from the vertical column with each other:

$$B = \frac{12}{4} = 3 \text{ or } B = \frac{4}{12} = \frac{1}{3}$$

So, B also has two values possible, $B = (3, \frac{1}{3})$

- So the number in the center can be obtained by combinations like AB, $A^2 + B^2$, A + B, A − B, ..., and so on. Since division is *non-commutative*, we can arrive at more values of the number in the center. Here are a few examples:

A + B:

$$12 + 3 = 15$$

$$12 + \frac{1}{3} = \frac{12 \times 3 + 1}{3} = \frac{36 + 1}{3} = \frac{37}{3}$$

$$\frac{1}{12} + 3 = \frac{1 + 3 \times 12}{12} = \frac{1 + 36}{12} = \frac{37}{12}$$

$$\frac{1}{12} + \frac{1}{3} = \frac{1 + 4}{12} = \frac{5}{12}$$

Hence, there are four possible values of A + B: $(15, \frac{37}{3}, \frac{37}{12}, \frac{5}{12})$

AB:

$$12 \times 3 = 36$$

$$12 \times \frac{1}{3} = 4$$

$$\frac{1}{12} \times 3 = \frac{1}{4}$$

$$\frac{1}{12} \times \frac{1}{3} = \frac{1}{36}$$

Hence, there are four possible values of AB: $(36, 4, \frac{1}{4}, \frac{1}{36})$. Since division and subtraction are non-commutative, the results will vary if the order of these arithmetic operations is changed. Please refer to section **2.2.4.4 Subtraction and division are non-commutative** (p. 48) for further discussions.

(v) Sum of squares: The number in the center is the sum of squares of each of the surrounding numbers.

As shown in the figure on the right side, the number in the center is:

$$2^2 + 4^2 + 3^2 + 2^2 = 4 + 16 + 9 + 4 = 33$$

A Number Cross

(vi) Sum of cubes: The number in the center is the sum of the cubes of each of the surrounding numbers.

As shown in the figure on the right side, the number in the center is:

$$3^3 + 2^3 + 1^3 + 4^3 = 27 + 8 + 1 + 64 = 100$$

A Number Cross

(vii) Arithmetic Mean: An *arithmetic mean* is the average of a list of numbers. It is calculated by dividing the sum of numbers by the count of the numbers in the list. In a *cross*, we have four surrounding numbers, so we can obtain the number in the center by adding up the four numbers and dividing the total by four.

As shown in the figure on the right, the number in the center is $\frac{5+1+8+6}{4} = \frac{20}{4} = 5$

A Number Cross

2.2.3.3 Calculate the number in the center using A and B

In the discussion for this section, we assume two numbers A and B as described below:

✔ Apply some arithmetic operations on the two numbers in the horizontal row and let us call the result A.

✔ Apply some arithmetic operations on the two numbers in the vertical column and let us call the result B.

✔ Depending on the problem, once we have A and B, the number in the center can be obtained by using one of the following patterns:

- AB,

- $A + B, A - B, B - A$

- $\dfrac{A}{B}, \dfrac{B}{A}$

- $A^2 + B^2$, etc.

🐎 **Rule of thumb: Start with the simple arithmetic operations**

In general, you should try using the simplest logic first, and if that does not work, then try another logic. The simplest logic is always preferred. Here the simple logic is one that uses simple arithmetic operations.

2.2.4.4 Subtraction and division are non-commutative

✔ A *binary operation* is *commutative* if changing the *order of operation* does not change the result. Arithmetic operations like addition and multiplication are *commutative*. This is obvious as $1 + 5 = 5 + 1$ or $2 \times 7 = 3 \times 7$. On the other hand, division and subtraction are *non-commutative* as $5 - 3 \neq 3 - 5$ or $\frac{1}{2} \neq \frac{2}{1}$. Understanding this concept crucial in finding the right answer for a *Number Cross problem*, where sometimes the *smaller* (or *larger*) *number* is subtracted from the larger (or smaller) number to determine what the number in the center is. Similarly, the number in the center can be determined by dividing the smaller (or larger) number by the larger (or smaller) number.

✔ In a cross, we have four numbers, and we have seen before how to yield two numbers out of these four numbers. For the discussion here, we will refer to these two newly generated numbers as *"the number on the left"* and *"the number on the right."* Continuing on, the order in which subtraction or division is performed could be positional, such as *subtracting the number on the left from the number on the right and vice versa* or *dividing the number on the left by the number on the right and vice versa*. it is obvious that the number in the center will vary depending on the order of subtraction or division.

2.2.3 Solving Steps

As we have already mentioned earlier, the number in the center is obtained by applying some combinations of arithmetic operations on the four surrounding numbers. In the discussion below, we will use the term "logic" to refer to such combinations of arithmetic operations. We will follow the steps laid down below to solve problems from this chapter:

✔ Review the problem by analyzing the first two crosses. Choose one of the first two crosses, and apply various arithmetic operations to the surrounding numbers in the cross. Find the appropriate combination of arithmetic operations that yields the number in the center of the chosen cross.

✔ Now apply the same logic on the other cross to confirm if the selected combination of arithmetic operations would yield the same number that is present in the center of the cross.

✔ Once we find a logic that is valid for the first two crosses, apply the same logic to find the missing in the third cross.

2.2.4 Solving the Problems

Now that we understand the basics behind a *Number Cross*, let us try solving the first problem from this chapter.

Problem (a)

Figure 2.3: Problem (a)

Let us try to solve this problem by attempting a few *arithmetic operations* to see which one works.

- **Addition**:

    The first cross: 3 + 1 + 3 + 2 = 9. This does not equal the number 18 at the center.

- **Multiplication**:

    The first cross: 3 × 1 × 3 × 2 = 18. This equals the number 18 at the center.

    The second cross: 3 × 2 × 2 × 3 = 36. This equals the number 36 at the center.

Since we have found the logic that works with the first two crosses, we can apply that same logic for the third cross to determine the missing number.

The third cross:

The number in the center is 1 × 2 × 2 × 3 = 12, and this is the answer.

Now that we have found the answer for problem (a), let us try to solve problem (b).

Problem (b)

Figure 2.4: Problem (b)

Let us try a few arithmetic operations as listed below:

- **Addition**:

    The first cross: 1 + 2 + 3 + 4 = 10. It does not equal number 24 in the center.

- **Multiplication**:

    The first cross: 1 × 2 × 3 × 4 = 24. This equals the number 24 at the center.

    The second cross: 3 × 3 × 9 × 4 = 324. This does not equal the number 84 at the center.

- **Mixed Arithmetic Operations:** We will try various combinations of arithmetic operations to calculate the number in the center, and select the appropriate combination, which yields the same number that is present in the center of the cross.

The first cross:

Add the numbers in the horizontal row $(A_1) = 1 + 3 = 4$

Add the numbers in the vertical column $(B_1) = 2 + 4 = 6$

The second cross:

Add the numbers in the horizontal row $(A_2) = 3 + 4 = 7$

Add the numbers in the vertical column $(B_2) = 3 + 9 = 12$

Let us now try the following combinations to see which one works in finding the answer:

$A + B$:

The first cross: $A_1 + B_1 = 4 + 6 = 10$. This result does not equal the number 24 in the center.

$A - B$:

The first cross: $A_1 - B_1 = 4 - 6 = -2$. This result does not equal the number 24 in the center.

AB:

The first cross: $A_1B_1 = 4 \times 6 = 24$. This result equals the number 24 in the center.

The second cross: $A_2B_2 = 7 \times 12 = 84$. This result equals the number 84 in the center.

We have found the logic that works for the first two crosses, so we can apply that same logic to the third cross to determine the missing number in the third cross as shown below.

The third cross:

Add the numbers in the horizontal row $(A_3) = 2 + 4 = 6$

Add the numbers in the vertical column $(B_3) = 5 - 2 = 3$

Hence, the answer is $A_3B_3 = 6 \times 3 = 18$

Please note that in this case, we found the answer rather quickly, but in most cases, you will need to try more combinations to see which one works for a given problem. In the above section, we discussed the various types of logic and methods that would be useful in determining the correct answer. We have already found the answers to both the problems, which we will show in the next section.

2.3 Answer

The answers to the problems are shown below:

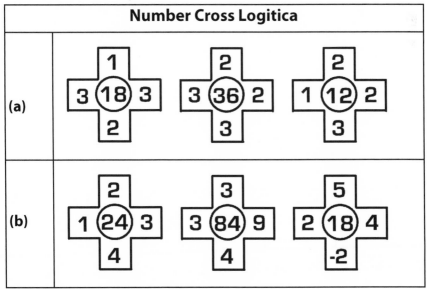

Figure 2.5: Answers

2.4 Summary

As we have discussed in this chapter, we use the four surrounding numbers in the cross to determine the number in the center of the cross. Hence, the *Number Cross* problems are more challenging than the *Number Box* problems. Since both use *arithmetic operations* to determine the missing number, it is easy to observe similarities in the methods used in finding the missing number.

We used two problems in this chapter to discuss the fundamentals behind a *Number Cross*. We followed a step-by-step approach to determine the missing number in the center. We also discussed some common arithmetic operations that can be used to determine the number in the center. However, you may find that there are even more combinations of these arithmetic operations possible that can be used in solving *Number Cross problems*. For example, the number in the center can be determined by using various combinations of arithmetic

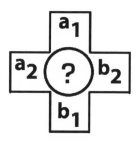

A Number Cross

operations like $\dfrac{a_1}{b_1} + \dfrac{a_2}{b_2}$, $a_1b_1 + a_2b_2$, $(a_1 + b_1)(a_1 - b_1)$, and etc.

It is recommended that readers practice with the problems provided in the *Exercise Section* to achieve a better understanding of solving *Number Cross* problems.

2.5 Exercise

Find the missing numbers in the problems below:

Level-I

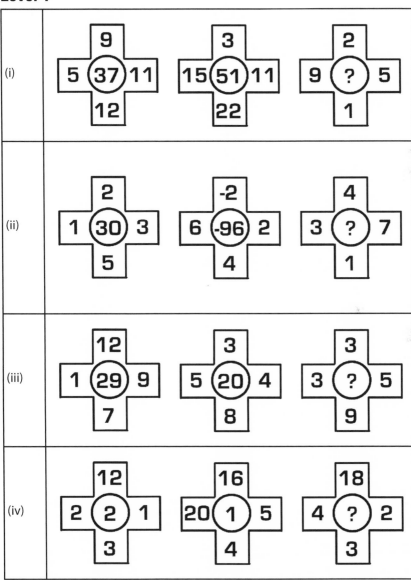

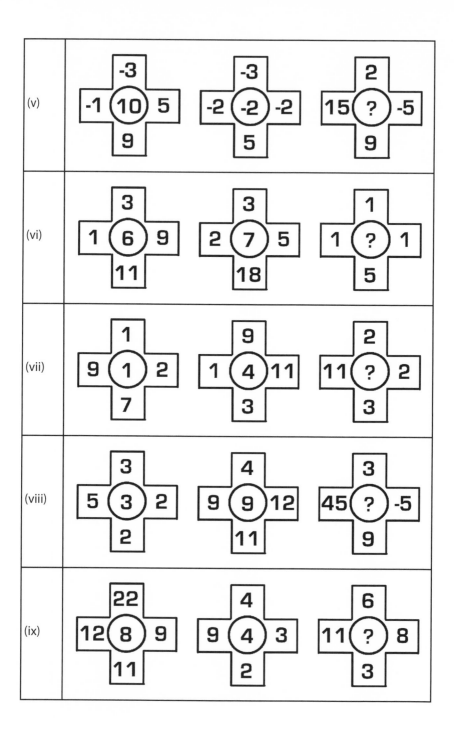

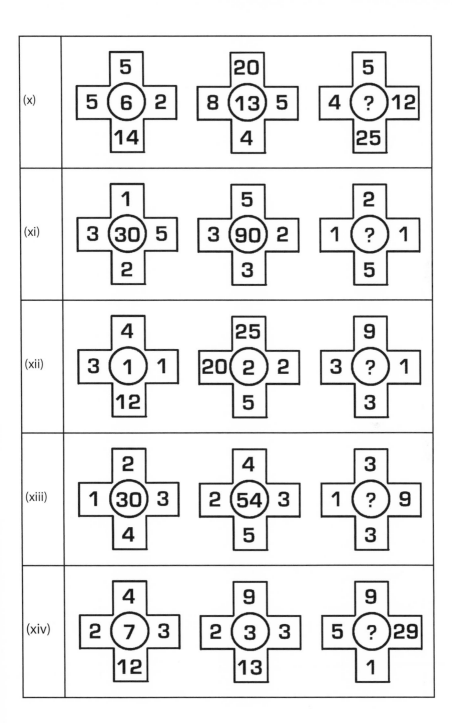

(xv)	20 / 15 (15) 12 / 13	5 / 11 (7) 9 / 3	20 / 30 (?) 10 / 12
(xvi)	1 / 1 (22) 11 / 9	4 / 2 (22) 5 / 11	3 / 5 (?) 4 / 11
(xvii)	1 / 1 (93) 3 / 4	3 / 1 (56) 1 / 3	3 / 1 (?) 3 / 2
(xviii)	32 / 11 (25) 46 / 11	10 / -5 (40) 25 / 30	20 / -5 (?) 10 / 11
(xix)	1 / -2 (84) 6 / -5	5 / 3 (25) 6 / -7	-3 / 4 (?) 2 / 2

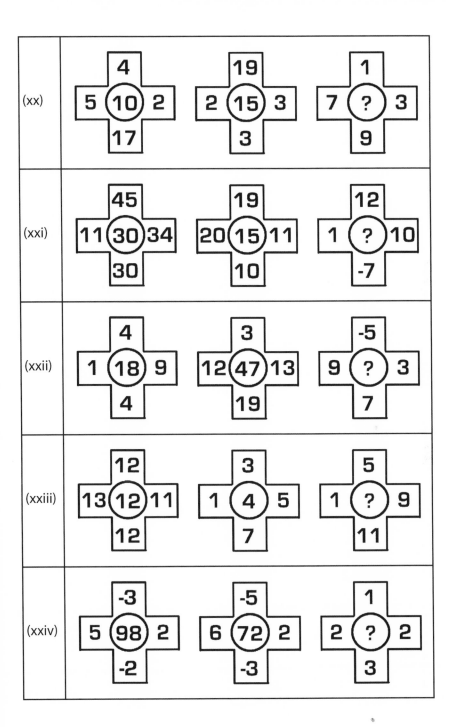

(xx)	4 5 (10) 2 17	19 2 (15) 3 3	1 7 (?) 3 9
(xxi)	45 11 (30) 34 30	19 20 (15) 11 10	12 1 (?) 10 -7
(xxii)	4 1 (18) 9 4	3 12 (47) 13 19	-5 9 (?) 3 7
(xxiii)	12 13 (12) 11 12	3 1 (4) 5 7	5 1 (?) 9 11
(xxiv)	-3 5 (98) 2 -2	-5 6 (72) 2 -3	1 2 (?) 2 3

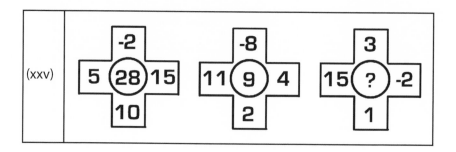

Level-II

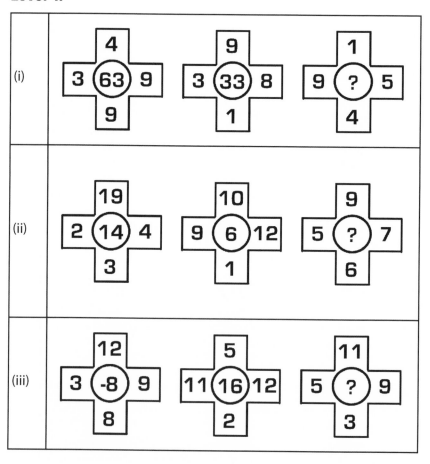

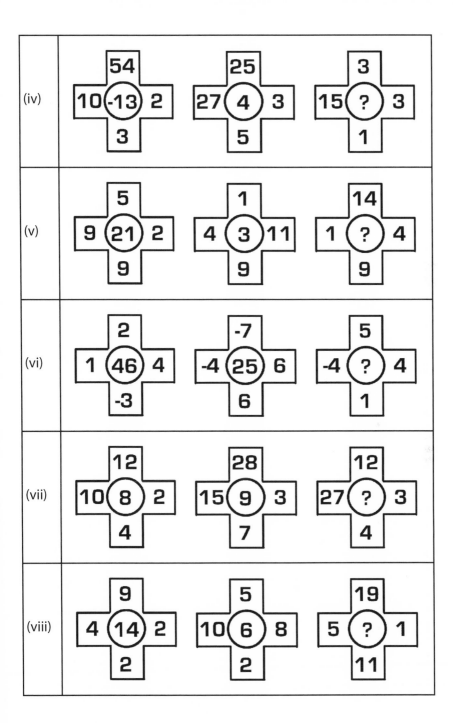

(iv)	54 / 10 (-13) 2 / 3 25 / 27 (4) 3 / 5 3 / 15 (?) 3 / 1
(v)	5 / 9 (21) 2 / 9 1 / 4 (3) 11 / 9 14 / 1 (?) 4 / 9
(vi)	2 / 1 (46) 4 / -3 -7 / -4 (25) 6 / 6 5 / -4 (?) 4 / 1
(vii)	12 / 10 (8) 2 / 4 28 / 15 (9) 3 / 7 12 / 27 (?) 3 / 4
(viii)	9 / 4 (14) 2 / 2 5 / 10 (6) 8 / 2 19 / 5 (?) 1 / 11

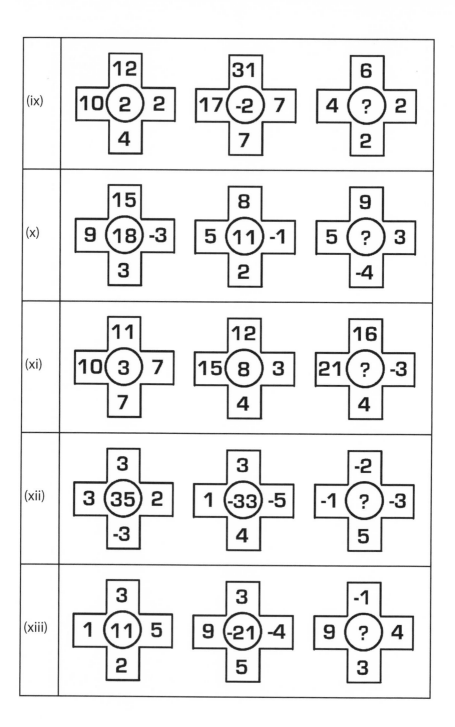

(ix)		
(x)		
(xi)		
(xii)		
(xiii)		

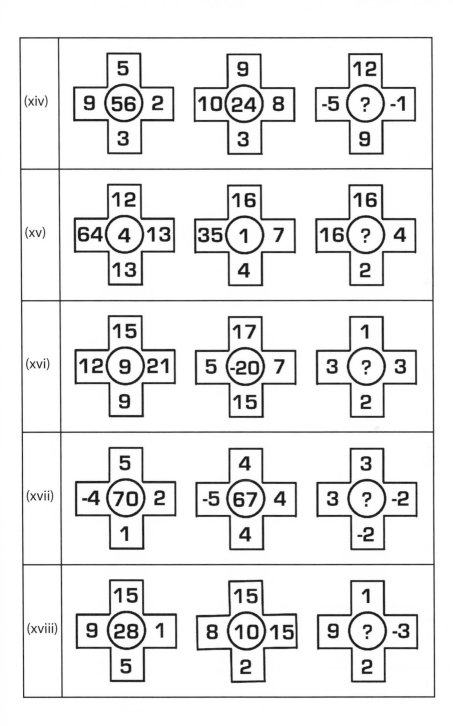

(xiv)	5 / 9 (56) 2 / 3	9 / 10 (24) 8 / 3	12 / -5 (?) -1 / 9
(xv)	12 / 64 (4) 13 / 13	16 / 35 (1) 7 / 4	16 / 16 (?) 4 / 2
(xvi)	15 / 12 (9) 21 / 9	17 / 5 (-20) 7 / 15	1 / 3 (?) 3 / 2
(xvii)	5 / -4 (70) 2 / 1	4 / -5 (67) 4 / 4	3 / 3 (?) -2 / -2
(xviii)	15 / 9 (28) 1 / 5	15 / 8 (10) 15 / 2	1 / 9 (?) -3 / 2

(xix)	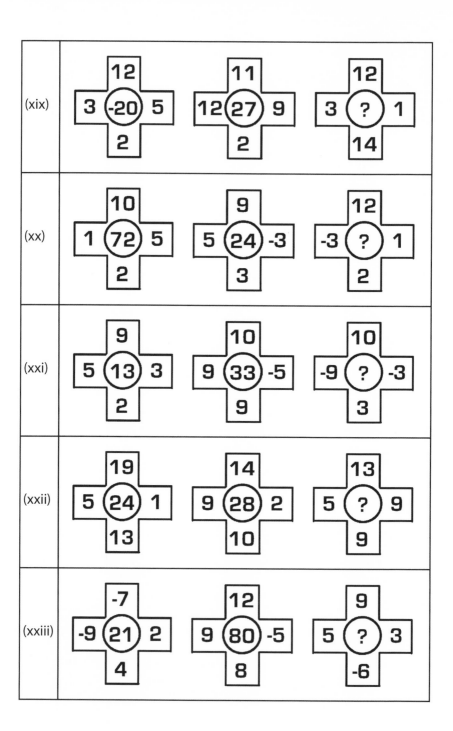
(xx)	
(xxi)	
(xxii)	
(xxiii)	

(xxiv)	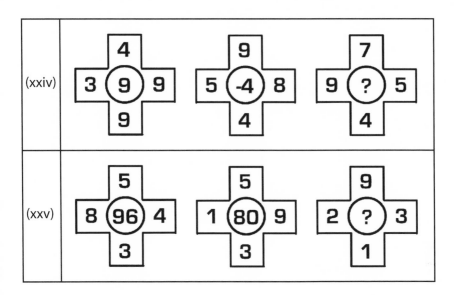
(xxv)	

Chapter 3

Marbles in a Box

Keywords: Linear Equations.

3.1 Challenge

Solve the following *"Marbles in a Box"* problems.

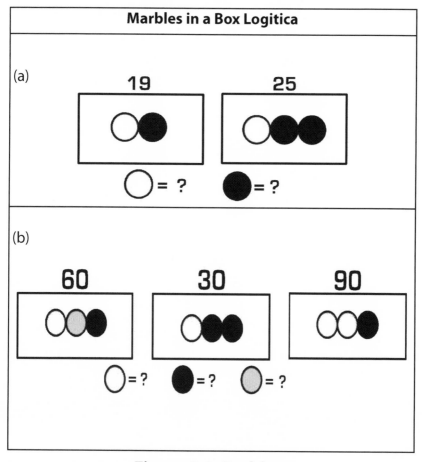

Figure 3.1: Problems

We are going to learn how to derive and use a set of *linear equations*[1] and how to use them for solving problems in this chapter. Let us first define what we mean by "*Marbles in a Box.*"

♛ Definition - Marbles in a Box

The problems in this chapter consist of **boxes containing marbles** of different colors (black, white, or gray). Each type of colored marbles is assigned a numerical value. Also, each individual box is labeled with a number, which is the sum of the numerical values of the marbles in that box. The numerical value is only assigned to the marbles, whereas the box itself does not contribute to the number displayed on it.

Objective: In this chapter, we need to find the missing numerical value for each type of colored marble based on the given information.

3.2 Strategy

In this section, we will give you an overview of the problems from this chapter. We will subsequently solve the problems after doing a thorough analysis.

3.2.1 Overview

The two problems in the chapter contain colored marbles in boxes. Based on this information, we need to find the missing number. Before we solve this problem, let us review a simple example of *"Marbles in a Box"* in the *Analysis Section*.

---

1.   Refer to **Appendix-D** for more details about *linear equations*.

3.2.2 Analysis

Before we solve the problems from this chapter, consider an example of *"Marbles in a Box,"* as shown on the right. As shown in the figure, we have a box with 2 *white marbles* and 1 *black marble*, and we need to find the value of the box. We can find the value of the box as shown below:

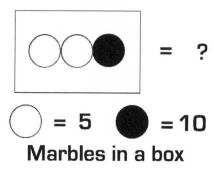

Marbles in a box

The value of *white marble* = 5

The value of *black marble* = 10

The box contains 2 *white marbles* and 1 *black marble*. As specified before, the value of the box is the sum of the numerical values of the marbles in the box. Hence, the value of the box = 2 × value of white marble + 1 × value of black marble = 2 × 5 + 1 × 10 = 10 + 10 = 20. This simple example demonstrates how the value of the box can be calculated using the marbles it contains. However, the problems in this chapter are slightly different from the problem we just discussed. The problems in this chapter have numbers given on the box, and we need to find the value of each type of colored marbles. But the simple example discussed here will help us to build the concept needed for solving problems from this chapter.

3.2.3 Solving Steps

We can follow the steps as described below to solve problems involving *"Marble in a Box"*:

✔ Review the problem by analyzing the boxes. Each box contains a unique combination of colored marbles.

✔ Each colored marble represents a numerical value, and the sum of the numerical values of marbles in each box is labeled on the box. So, we can write equations using this information.

✔ Once we have the required number of equations, we can solve them to find the numerical value of each marble.

3.2.4 Solving the Problems

Let us first solve problem (a).

Problem (a)

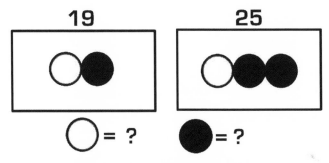

Figure 3.2: Problem (a)

In this problem, we have two boxes labeled with numerical values, and we need to find the value of each type of colored marbles. Each of the boxes provided in figure-3.1 has a unique combination of black and white marble(s) as described below:

• The first box has 1 *white marble* and 1 *black marble*. The sum of this combination of marbles is 19.

• The second box has 1 *white marble* and 2 *black marbles*. The sum of this combination of marbles is 25.

• Using the above information, we need to determine the numerical value for each of the white and black marbles.

Based on the above observations, we can derive a set of *linear equations* by using the combination of marbles in each box. We assume the following symbols to derive the required equations:

w = The value of white marble

b = The value of black marble

From the first box   : $w + b = 19$ Equation (1)

From the second box : $w + 2b = 25$ Equation (2)

Now we have a pair of equations and two variables: w and b. We can solve these equations by following a step-by-step approach as shown below:

✔ **Step 1**: To find b, we need to subtract equation (1) from (2) as shown below:

$$w + 2b = 25 \qquad \text{[Equation (2)]}$$

$$w + b = 19 \qquad \text{[Equation (1)]}$$

Subtracting both sides of the equations:

$$(w + 2b) - (w + b) = 25 - 19$$

$$w + 2b - w - b = 6$$

$$b = 6$$

✔ **Step 2**: To find w, we can substitute the value of b in either equation (2) or (1). Let us take equation (1):

$$w + b = 19 \qquad \text{[Equation (1)]}$$

$$w + 6 = 19 \qquad \text{[Using } b = 6]$$

$$w = 19 - 6 = 13$$

The calculated values of marbles w and b are shown below:

$$w = 13$$

$$b = 6$$

✔ **Step 3**: This is the verification step, where we substitute $w = 13$, $b = 6$ in equations (1) and (2) to verify that the calculated values are correct:

Equation (1):

$$w + b = 13 + 6 = 19$$

Equation (2):

$$w + 2b = 13 + 2 \times 6 = 13 + 12 = 25$$

Since the above-calculated values match with the right-hand side of the respective equations, we conclude that the calculated values are correct. In this problem, we had white and black marbles, which were represented by two variables w and b. Therefore, we needed two independent equations[1] to find the value of each variable. Once we were able to calculate the value of each variable, then determining the numerical value for the combination of marbles in the third box was fairly straightforward. Let us continue our discussion by focusing our attention on problem (b).

Problem (b)

In this problem, we have three boxes labeled with numerical values, and we need to find the value of each type of colored marbles. Each of the boxes provided in figure-3.1 has a unique combination of white, black and gray marble(s) as described below:

- The first box has 1 *white marble,* 1 *black marble,* and 1 *gray marble.* The sum of this combination of marbles is 60.

- The second box has 1 *white marble* and 2 *black marbles.* The sum of this combination of marbles is 30.

- The third box has 2 *white marbles* and 1 *black marble.* The sum of this combination of marbles is 90.

---

1. Readers can refer to **Appendix-D** (p. 165) for more details about *independent equations.*

- Using the above information, we need to determine the numerical value for each of the white, black and gray marbles.

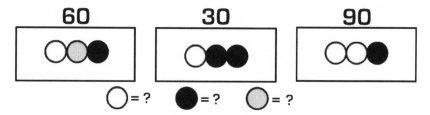

Figure 3.3: Problem (b)

Based on the above observations, we can derive a set of *linear equations* by using the combination of marbles in each box. We assume the following symbols to derive the required equations:

w = The value of white marble

b = The value of black marble

g = The value of gray marble

From the first box : $w + b + g = 60$ Equation (1)

From the second box : $w + 2b = 30$ Equation (2)

From the third box : $2w + b = 90$ Equation (3)

Now we have three equations and three variables: w, b, and g. We can solve these equations by following a step-by-step approach as shown below:

Step-1: If we multiply equation (2) by 2, we get a transformed equation (2a) as shown below:

$$w + 2b = 30 \qquad \text{[Equation (2)]}$$

Multiply both sides by 2:

$$2(w + 2b) = 2 \times 30$$

$$2w + 4b = 60 \qquad \text{[Equation (2a)]}$$

Step-2: We can now subtract equation (3) from (2a), we can find the value of w as shown below:

$$2w + 4b = 60 \qquad \text{[Equation (2a)]}$$

$$2w + b = 90 \qquad \text{[Equation (3)]}$$

Subtracting both sides of the equations:

$$(2w + 4b) - (2w + b) = 60 - 90$$

$$2w + 4b - 2w - b = -30$$

$$3b = -30$$

$$b = -\frac{30}{3}$$

$$b = -10$$

Step-2: We can use the value of b in either equation (2) or (3) to find the value of w. Let us take equation (2):

$$w + 2b = 30 \qquad \text{[Equation (2)]}$$

$$w + 2 \times (-10) = 30 \qquad \text{[Using } b = -10]$$

$$w - 20 = 30$$

$$w = 30 + 20$$

$$w = 50$$

Step-3: We can use the values of w and b in equation (1) to find the value of g:

$$w + b + g = 60 \qquad \text{[Equation (2)]}$$

$$50 - 10 + g = 60 \qquad \text{[Using } w = 50, \ b = -10]$$

$$40 + g = 60$$

$$g = 60 - 40$$

$$g = 20$$

The calculated values of marbles w, b and g are shown below:

$$w = 50, b = -10, g = 20$$

Step 4: This is the verification step, where we substitute $w = 50$, $b = -10$, and $g = 20$ in equations (1), (2), and (3) to verify that the calculated values are correct:

> **Equation (1):**
>
> $$w + b + g = 50 - 10 + 20 = 60$$
>
> **Equation (2):**
>
> $$w + 2b = 50 + 2 \times (-10) = 50 - 20 = 30$$
>
> **Equation (3):**
>
> $$2w + b = 2 \times 50 - 10 = 100 - 10 = 90$$

Since the above-calculated values match with the right-hand side of the respective equations, we conclude that the calculated values are correct.

We have already found the answers to both the problems, which we will show in the next section.

3.3 Answer

The answers to the problems are shown below:

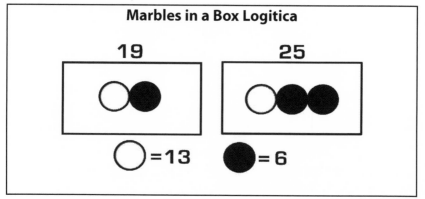

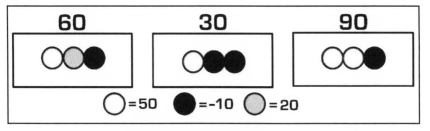

Figure 3.4: Answers

3.4 Summary

In this chapter, we solved two problems. Problem (a) contains two types of colored marbles (white and black), so we needed two *independent equations to solve* the problem. On the other hand, problem (b) contains three types of colored marbles (white, black, and gray). Therefore, we required three independent equations[1] to solve the problem.

In general, *to find values of N variables, you need N independent equations*. There are two ways of solving a set of *linear equations*: *elimination* and *substitution*[2]. In this chapter, whenever appropriate, we have preferred one over the other in order to solve the equations.

Readers are advised to try and solve the problems provided in the *Exercise Section* of this chapter in order to improve their understanding of concepts in this chapter.

---

1.   Refer to **Appendix-D** (p. 165) for further details on *independent equations*.

2.   Readers can refer to **Appendix-D** (p. 165) for more details on the two methods of solving linear equations: *elimination* and *substitution*.

3.5 Exercise

Find the missing numbers in the following problems:

Level-I

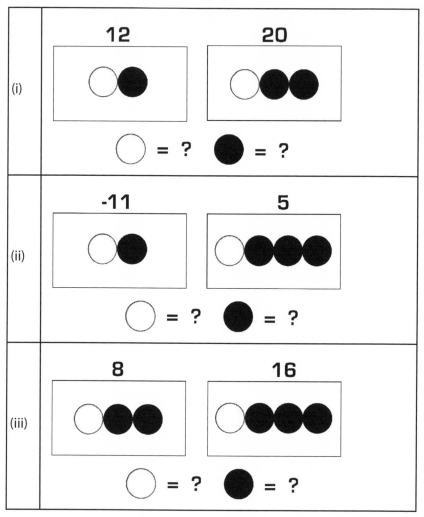

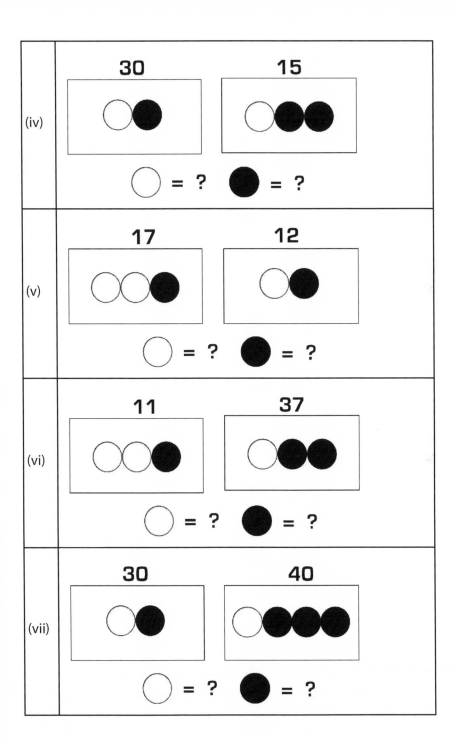

(iv)

30

15

◯ = ?   ⬤ = ?

(v)

17

12

◯ = ?   ⬤ = ?

(vi)

11

37

◯ = ?   ⬤ = ?

(vii)

30

40

◯ = ?   ⬤ = ?

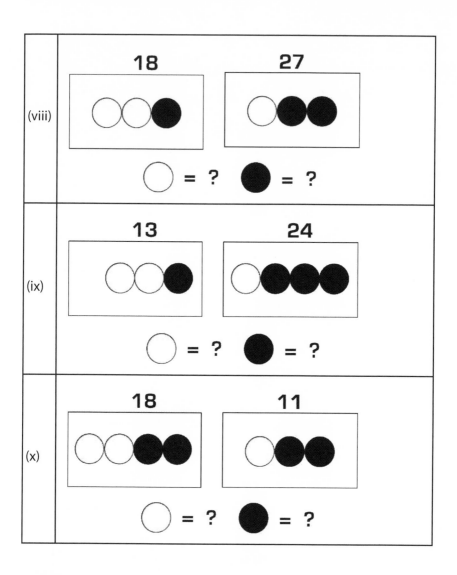

(viii)

18

27

○○● ○●●

○ = ? ● = ?

(ix)

13

24

○○● ○●●●

○ = ? ● = ?

(x)

18

11

○○●● ○●●

○ = ? ● = ?

Level-II

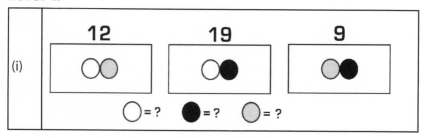

(i)

12

19

9

○◉ ○● ◉●

○ = ? ● = ? ◉ = ?

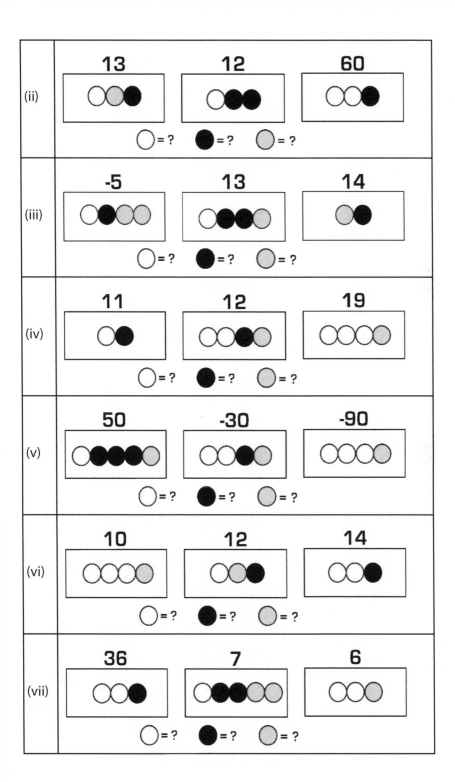

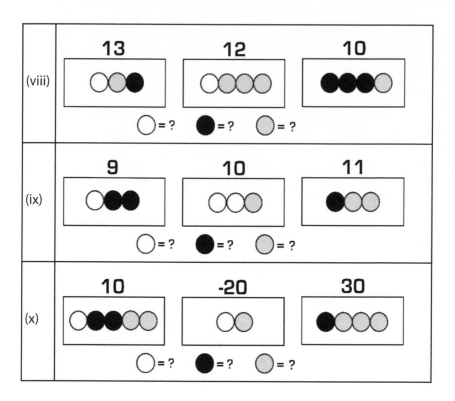

Chapter 4

Average Cell

Keywords: Linear Equations, Fractions, Arithmetic Mean.

4.1 Challenge

Solve the following *Average Cell* problems:

Average Cell Logitica
(a) Find the value of A. 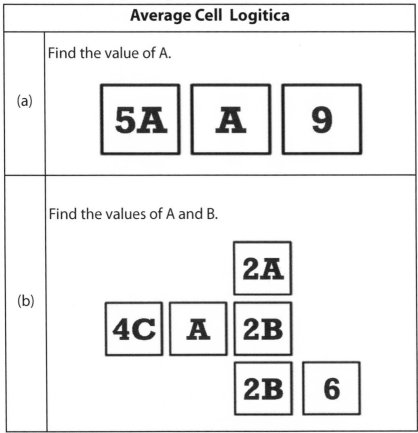
(b) Find the values of A and B.

Figure 4.1: Problems

In this chapter, we will learn how to derive a set of linear equations[1] in an interesting way, and use these equations to solve problems involving *Average Cells*. Let us first define what we mean by an *Average Cell*.

♛ Definition - Average Cell

In a cluster of cells, any cell that has more than one cell adjacent to it (above, below, left, or right), would be an **Average Cell**. Below are the required conditions for an *Average Cell*:

- The number indicated in the *Average Cell* is the average of all the numbers in the adjacent cells.
- We will use the number count of the adjacent cells as the divisor in calculating the average.
- Diagonal cells are not considered to be adjacent cells when solving problems in this chapter.

Objective: In this chapter, we aim to find the numerical value of each cell in a cluster of cells arranged in a particular way, similar to what is shown in figure-4.1.

4.2 Strategy

In this section, we will first give an overview of the problem involving *Average Cells*, which is followed by a thorough analysis of the topic. Later in the chapter, We will outline steps describing how to solve such type of problems.

4.2.1 Overview

To solve problems involving *Average Cells*, we first need to identify all the *Average Cells* in the problem. As defined earlier,

---

1. Refer to **Appendix-D** for more details about *linear equations*.

the value of an *Average Cell* is the average of the numbers in the surrounding cells. Once we identify all the *Average Cells,* we can write equations using the formula for calculating the average. We can subsequently use these equations to solve for the variables (e.g. A, B or C). Once we know the numerical values of each variable, we can complete each cell with its numerical value.

4.2.2 Analysis

To solve the problem, we first need to identify which cells should be selected for calculating the average. According to the definition, the number in an *Average Cell* is the average of the numbers in the cells adjacent to it.

An average is also termed as an "*arithmetic mean.*" We have discussed *arithmetic mean* when discussing *Number Box* problems in **Chapter-1 (p. 18)**. Let us revisit this quickly to refresh our understanding.

♛ Definition - Arithmetic Mean or Average

An *arithmetic mean* is the average of a list of numbers. It is defined as the sum of the numbers divided by the count of these numbers. Here are a few examples of how to calculate the average:

(i) The average of 7 and 8 $= \dfrac{7+8}{2} = \dfrac{15}{2}$

(ii) The average of 1, 2, and 3 $= \dfrac{1+2+3}{3} = \dfrac{6}{3} = 2$

(iii) If there are three unknown numbers represented by variables A, B, and C, then the average these variables is $\dfrac{A+B+C}{3}$.

In the next section, we will outline the steps to solve problems from the chapter.

4.2.3 Solving Steps

We can follow the steps as described below for solving a problem involving *Average Cells:*

✔ Identify all the *Average Cells* in the problem.

✔ Write an equation representing the value of each *Average Cell* as the average of its surrounding numbers. Repeat this step for all of the *Average Cells* in the problem

✔ Once we have the required number of equations, we can solve them to find the value of the variables in the equations. We can then use the value of variables to determine the value of each cell in the problem.

4.2.4 Solving the Problems

Let us use the steps laid out in the previous section to solve two problems from the chapter. Let us start with problem (a).

Problem (a)

To indicate which cells we have selected, and the variable inside that will be used in calculating the average, we have marked them with circles in figure-4.2. A *black circle* is used to highlight the variable inside the *Average Cell* in figure figure-4.2.

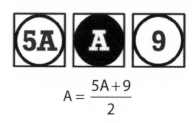

$$A = \frac{5A + 9}{2}$$

Figure 4.2: Problem (a)

As shown below in figure-4.2, we have only one *Average Cell*. In this problem, A is the *average* of 5A and 9 as shown below:

$$A = \frac{5A+9}{2}$$

$$2A = 5A + 9$$

$$5A - 2A = -9$$

$$3A = -9$$

$$A = -\frac{9}{3} = -3$$

Now that we have found the value of A, let us verify the answer using the definition of an *Average Cell*.

Value of the *Average Cell* = A $= -3$
$A = \dfrac{5A+9}{2}$
$= \dfrac{5 \times (-3)+9}{2}$
$= \dfrac{-15+9}{2}$
$= \dfrac{-6}{2}$
$= -3$
This proves the answer.

Figure 4.3: Verification Steps

While the above verification step seems trivial, we will see that this step is quite useful in verifying advanced problems like problem (b). We will summarize the answer in the *Answer Section* later in the chapter. Let us now solve the problem (b).

Problem (b)

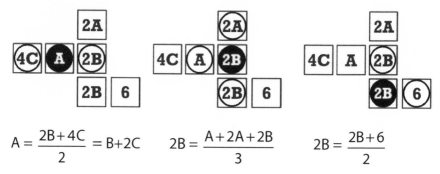

$$A = \frac{2B+4C}{2} = B+2C \qquad 2B = \frac{A+2A+2B}{3} \qquad 2B = \frac{2B+6}{2}$$

Figure 4.4: Average Cells

In figure-4.4 we have 3 images for each of the *Average Cells*. To indicate which cells we have selected, and the variable inside that will be used in calculating the average, we have marked them with circles in the images in figure-4.4. A black circle is used to highlight the variable inside the *Average Cell* in each of the images in figure-4.4.

As shown in figure-4.4, we can quickly identify three *Average Cells* in problem (b). Based on the figure, we can derive three equations as described below:

First image:

In this image, A is the *Average Cell,* and it is the average of the values of 2B and 4C as expressed in the equation below:

$$A = \frac{2B+4C}{2} = \frac{2(B+2C)}{2}$$

$$A = B + 2C$$

$$A - B - 2C = 0 \qquad \text{Equation (1)}$$

Second image:

In this image, 2B is the *Average Cell*, and it is the average of A, 2A, and 2B as shown in the equation below:

$$\frac{A + 2A + 2B}{3} = 2B$$

$$\frac{3A + 2B}{3} = 2B$$

$$3A + 2B = 3 \times 2B = 6B$$

$$3A = 6B - 2B$$

$$3A = 4B$$

$$A = \frac{4B}{3} \qquad \text{Equation (2)}$$

Third image:

In this image, 2B is the *Average Cell*, which is the average of 2B and 6 as shown in the equation below:

$$\frac{2B + 6}{2} = 2B$$

$$2B + 6 = 2 \times 2B = 4B$$

$$4B - 2B = 6$$

$$2B = 6$$

$$B = \frac{6}{2} = 3 \qquad \text{Equation (3)}$$

We have listed down the derived equations below for legibility:

$$A - B - 2C = 0 \qquad \text{Equation (1)}$$

$$A = \frac{4B}{3} \qquad \text{Equation (2)}$$

$$B = 3 \qquad \text{Equation (3)}$$

Now we will follow a step-by-step method to find the values of A,

B, and C as shown below:

Step 1: We have already found the value of B, we can use it in equation (2) to find the value of A as shown below:

$$A = \frac{4B}{3} \qquad \text{[Equation (2)]}$$

$$A = \frac{4 \times 3}{3} \qquad \text{[Using B = 3 from equation(3)]}$$

$$A = \frac{12}{3} = 4$$

Step 2: We can use the value of A and B in equation (1) to find the value of C as shown below:

$$A - B - 2C = 0 \qquad \text{[Equation (1)]}$$

$$4 - 3 - 2C = 0 \qquad \text{[Using A = 4, B = 3]}$$

$$2C = 1$$

$$C = \frac{1}{2}$$

Step 3: Now that we have calculated the values of all the variables, we can summarize them below:

$$A = 4$$

$$B = 3$$

$$C = \frac{1}{2}$$

Step 6: This is the verification step, where we will verify the calculated values of A, B, and C using the definition of the *Average Cell*. We will use figure-4.4 to verify our answers. If we have derived and solved the equations correctly, the numerical value of each *Average Cell* will be equal to the average of the adjacent cells. This step is shown in the table in figure-4.5.

First image	Second image	Third image
Value of the *Average Cell* **= A** $A = 4$	**Value of the** *Average Cell* **= 2B** $2B = 6$	**Value of the** *Average Cell* **= 2B** $2B = 6$
$A = \dfrac{2B+4C}{2}$ $= \dfrac{2\times3+4\times\dfrac{1}{2}}{2}$ $= \dfrac{6+2}{2} = \dfrac{8}{2} = 4$ This proves the answer.	$2B = \dfrac{A+2A+2B}{3}$ $= \dfrac{3A+2B}{3}$ $= \dfrac{3\times4+2\times3}{3}$ $= \dfrac{12+6}{3} = \dfrac{18}{3} = 6$ This proves the answer.	$2B = \dfrac{2B+6}{2}$ $= \dfrac{2\times3+6}{2}$ $= \dfrac{6+6}{2} = \dfrac{12}{2} = 6$ This proves the answer.

Figure 4.5: Verification steps

Explanation of the table in figure-4.5

In the table above, we have the following information:

- The first column refers to the first image in figure-4.4, in which A is the *Average Cell.*

- The second column refers to the second image in figure-4.4, in which 2B is the *Average Cell.*

- The third column refers to the third image in figure-4.4, in which 2B is the *Average Cell.*

- In each of the columns, we have shown the steps to verify the answer.

As shown in the table above in figure-4.5, we have verified that the value of each *Average Cell* is indeed the average of its adjacent cells. We now can summarize the verified answer in the next section.

4.3 Answers

The answers to the problems are shown below:

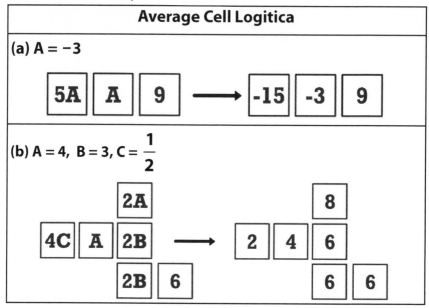

Figure 4.6: Answers

4.4 Summary

We started the chapter with problem (a). While solving this problem, we calculated the average of the adjacent cells and followed a step-by-step approach to derive and solve a set of *linear equations*. In the end, we used the definition of the *Average Cell* to verify our calculations. In problem (b), we discussed another problem with a different cluster of cells. The approach taken to determine and verify the values in this problem was similar to that of problem (a). The two problems discussed in the chapter should help to build a strong foundation for solving this type of problems. There are two ways of solving a set of linear equations: *elimination and substitution*[1]. In this chapter, whenever appropriate, we have preferred one over the other in order to solve the equations.

Readers are advised to try and solve the problems provided in the *Exercise Section* of this chapter in order to improve their understanding of concepts in this chapter.

---

1. Readers can refer to **Appendix-D** (p. 165) for more details about *these methods* for solving *linear equations.*

4.5 Exercise

Level-I

Solve the following *Average Cell* problems.

(i)	(ii)
$4A$ A 10	$6A$ $-A$ -24

(iii)	(iv)
A A 12	A $A+8$ 12

(v)	(vi)
$A-5$ $A+5$ 12	$6A$ $2A$ 10

(vii)	(viii)
A $A+8$ 20	$3A$ $4A$ $A-48$

(ix)	(x)
$A+8$ $A-2$ $2A$	$5A$ $3A$ 21

Level-II

Solve the following *Average Cell* problems.

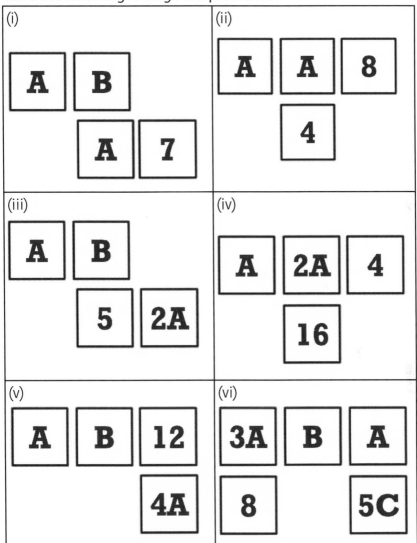

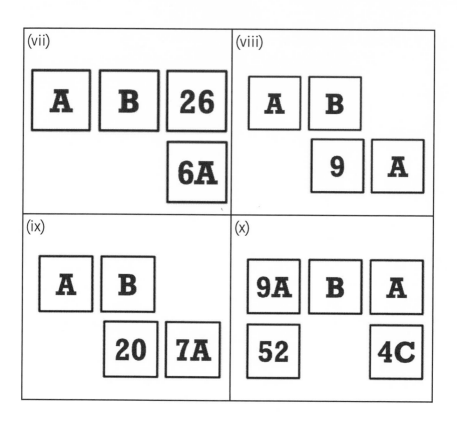

Chapter 5

Wisgo Number Tile

Keywords: Stimulating the left and right sides of the brain.

5.1 Challenge

Solve the following *Wisgo Number Tiles problems*:

Wisgo Number Tiles Logitica
Find the missing numbers.

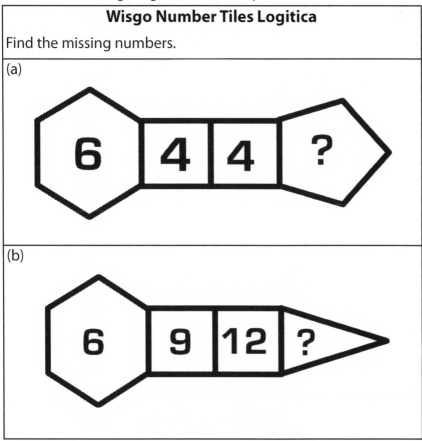

Figure 5.1: Problems

In this chapter, we need to find the missing number in a particular arrangement of tiles, each of which is called a *Wisgo Number Tile*. *Wisgo Number Tile* belongs to a family of problems called "*Wisgo Family*." Let us start this chapter by defining what we mean by *Wisgo Family* and *Wisgo Number Tile*.

♛ Definition - Wisgo Family

A problem in **Wisgo Family** is composed of a variety of *geometrical shapes* and *numbers*. To solve this category of problems, we need to analyze the shapes and numbers and figure out how they are related. In this chapter, we will discuss the simplest type of problem from the *Wisgo Family*, which is the *Wisgo Number Tile*. The discussion on other types of problems from the *Wisgo Family* is beyond the scope of this book.

♛ Definition - Wisgo Number Tile

A **Wisgo Number Tile** is a tile with a number on it. A tile can have different types of sides: *enclosed*, *shared*, and *non-shared* sides. The number on the tile depends on the type and count of sides included in the counting. For example, in a problem, the number on a tile may be related to the count of shared-sides of the tile. More details on this will be discussed later in the chapter.

Objective: The problems in this chapter are composed of different tiles joined together, similar to what is shown in figure-5.1. In a problem involving *Wisgo Number Tiles*, one of the tiles is missing its value. The objective of the problem is to determine the missing value.

♖ **Note**: For brevity, throughout this chapter, we will refer to a *Wisgo Number Tile* as a "*tile*" or a "*Wisgo Tile*."

5.2 Strategy

In this section, we will give an overview of *Wisgo Number Tile*. Subsequently, we will do a thorough analysis of the topic and define steps for solving such problems. Once we have acquired sufficient knowledge on the topic, we will solve the two problems from the chapter.

5.2.1 Overview

As mentioned before, the objective of a *Wisgo Number tiles* problem is to determine the missing number. Consider an example in Figure-5.2.

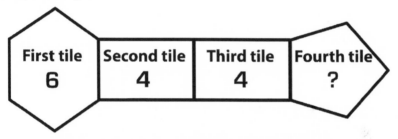

Figure 5.2: Four Wisgo Number Tiles

Figure-5.2 contains four tiles, and the numbers on the first three tiles are 6, 4, and 4. Based on this information, we need to determine the missing value on the fourth tile. Later in the chapter, we will see how to analyze and solve this problem.

5.2.2 Analysis

First, we need to understand the fundamentals behind a *Wisgo Number Tile*. The number on a tile is related to its sides in two different ways as described below:

i. The sides of a tile: The sides of a tile refer to its *enclosing sides*.

To solve a problem involving *Wisgo Tiles*, we should first try to find the relationship between the numerical value on the tile and the number of sides the tile has. For example, if a tile has 5 sides, the number on the tile could be related to number 5.

ii. Shared vs. non-shared sides: In some problems, *counting how many enclosing sides a tile has is not sufficient.* We also need to check whether the tile has any sides in common with one or more other tiles. In other words, we need to explore whether the numerical value on the tile is related to the number of *shared* or *non-shared sides* or *both*. The shared side of a tile is the side that is in common with another tile(s) connected to it, whereas the non-shared side is the side that does not have any tile connected to it. Let us discuss with an example as shown in figure-5.3.

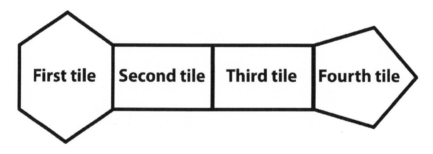

Figure 5.3: Wisgo Number Tiles

Referring to the tiles in figure-5.3:

- ✔ The first tile has 6 sides and shares its right side with the second tile. It has no tile on its left.
- ✔ The second tile has 4 sides. It shares its left side with the first tile and right side with the third tile.
- ✔ The third tile has 4 sides. It shares its left side with the second tile and right side with the fourth tile.
- ✔ The fourth tile has 5 sides and shares its left side with the

third tile, and it has no tile on its right.

According to the definition of *Wisgo Number Tile*, the number on the tile depends on the types of sides included in counting the number of sides. Let us review this with a few examples. In the following discussion, we will explain how to count the number of sides in a tile. We have used a *solid black line* for the sides that will be included in this count, and a *dotted gray line* for the sides that will be excluded from the count. Consider the following examples:

✔ When the number on the tile is the same as the number of sides of the tile:

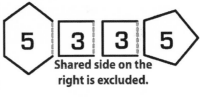

- The number on the first tile will be 6.
- The number on the second tile will be 4.
- The number on the third tile will be 4.
- The number on the fourth tile will be 5.

✔ When the number on the tile is the same as the number of sides of the tile, excluding the shared side on the right:

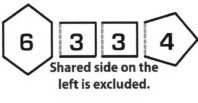

Shared side on the right is excluded.

- The number on the first tile will be 5.
- The number on the second tile will be 3.
- The number on the third tile will be 3.
- The number on the fourth tile will be 5.

✔ When the number on the tile is the same as the number of sides of the tile, excluding the shared side on the left:

Shared side on the left is excluded.

- The number on the first tile will be 6.

- The number on the second tile will be 3.
- The number on the third tile will be 3.
- The number on the fourth tile will be 4.

✔ When the number on the tile is the same as the number of non-shared sides:

All shared sides are excluded.

- The number on the first tile will be 5.
- The number on the second tile will be 2.
- The number on the third tile will be 2.
- The number on the fourth tile will be 4.

✔ When the number on the tile is the same as the number of shared sides:

Only shared sides are included.

- The number on the first tile will be 1.
- The number on the second tile will be 2.
- The number on the third tile will be 2.
- The number on the fourth tile will be 1.

In the examples above, depending on how sides were calculated, the number on a tile was the same as the number of sides (*enclosing*, *shared* or *non-shared*) the tile had. However, in a different problem, the number on the tile may be related to or derived from the number of sides. For example, if the number of sides of a tile is counted as 6, then the number on the tile could be 21, which is the sum of the first 6 natural numbers, i.e., $1 + 2 + 3 + 4 + 5 + 6 = 21$. Along the same lines, in a different problem, the number on the tile may be obtained by applying different arithmetic operations on the numbers obtained by counting sides.

5.2.3 Solving Steps

We can follow the steps as described below to solve a *Wisgo Tiles* problem:

✔ Review the problem by observing the shape and number on each of the tiles.

✔ Observe how numbers are changing on the tiles. Do you see any pattern or logic?

✔ Try to find a relation between numbers on the tile and count of the sides, where sides could be of *enclosed*, *shared*, and *non-shared* types. Use this relation to form a logic that relates the number on a tile to the count and type of sides included in the count.

✔ Verify that the logic is valid for all the known numbers in the problem.

✔ Once we discover a logic that is consistent with all the tiles, we can use the same logic to find the missing number.

5.2.4 Solving the Problems

Now that we have understood the basics of *Wisgo Number Tiles*, let us solve our first problem.

Problem (a)

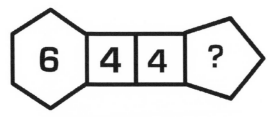

Figure 5.4: Problem (a)

We can make the following observations by analyzing the

number of sides of the tile and the number indicated on it:

- The number on the first tile is 6, which equals the number of sides of the first tile.

- The number on the second tile is 4, which equals the number of sides of the second tile.

- The number on the third tile is 4, which equals the number of sides of the third tile.

- Following the same logic, the fourth tile has 5 sides, which means the number on this tile will be 5. This is the answer to the problem, which we will summarize in the *Answer Section* of the chapter.

This was a simple problem, as the number on the tile was the same as the number of enclosed sides each tile had. Let us continue our discussion with problem (b)

Problem (b)

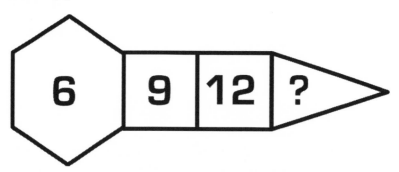

Figure 5.5: Problem (b)

The known numbers on the tiles are 6, 9, and 12. Since there is no obvious pattern in the numbers on the tiles like we saw in the previous problem, the problem needs to be analyzed further. In the section below, we have filled in the tile with black to highlight the relevant tile under discussion.

✔ **Step 1**: As shown in the figure on the right, the number on this tile is 6. The first tile has 6 sides. Does it mean that the number on the tile equals the number of sides the tile has? We need to verify if this logic is valid for the next tile.

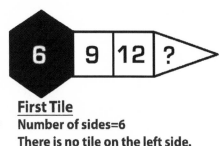

First Tile
Number of sides=6
There is no tile on the left side.

✔ **Step 2**: The second tile has 4 sides. However, the number on the second tile is 9, which is not equal to the number of sides of the second tile. To solve this problem, let us take a closer look at the numbers on the other tiles. We notice that as we move from left to right, the *numbers on the tiles are increasing* (i.e., 6, 9, and 12). Let us calculate the difference between these numbers to see if we discover any patterns in the increments:

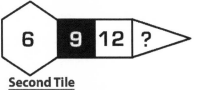

Second Tile
Number of sides=4
Number of sides excluding the left side=3

$$9 - 6 = 3$$
$$12 - 9 = 3$$

So the numbers are incrementing by 3. Did you notice that the count of sides of the second and third tile is 4? Could this mean that one of the shared sides between two tiles is not included in the count? Let us analyze this logic in the next step.

✔ **Step 3**: The third tile has 4 sides. As we have already seen in the previous step, the difference between the numbers on the second and third tiles is 3. Note that 3 is the same result we can get by excluding one of the shared sides in the count of the sides.

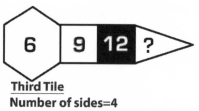

Third Tile
Number of sides=4
Number of sides excluding the left side=3

Hence, we can conclude that one of the shared sides between the tiles has been excluded when counting the number of sides. But which one, shared side on the left or shared side on the right? The clue lies in the first tile as described below:

(a) We know for sure that out of the left and right sides, one of the shared sides is excluded when counting the number of sides.

(b) Had the shared side on the right been excluded, the number on the first tile would have been 5, but it is not. Instead, the number on the first tile is 6.

Based on the above two points, we can conclude that the shared side on the left was not included in the count. Now that we have discovered a consistent logic that fits well with the number on each tile, we can apply the same logic to determine the missing number on the fourth tile.

✔ **Step 4**: The fourth tile has 3 sides, and so the number of sides excluding the shared side on the left is 2. Hence, to find the

Fourth Tile
Number of sides=3
Number of sides excluding the left side=2

number on the fourth tile, we just need to add 2 to the number on the previous tile: 12 + 2 = 14, and this is the answer to the problem.

Summary of steps

Let us summarize the key points from the above discussions:

- We count the number of sides of a tile excluding the shared side on the left.

- We then add the count to the number on the previous tile. This step is repeated for each of the tiles.

♜ **Note**: Since the first tile has 6 sides, and there is no tile on its left. Hence, the number on the first tile is 6.

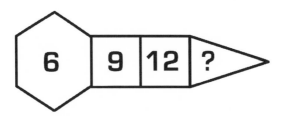

Figure 5.6: Problem (b)

Let us apply the above two steps for each of the tiles:

First tile:

The number of sides = 6 and there is no tile on its left.

The number on this tile = 6

Second tile:

The number of sides = 4

The number of sides excluding the shared side on the left = 3

The number on the previous tile = 6

The number on this tile = 6 + 3 = 9

Third tile:

The number of sides = 4

The number of sides excluding the shared side on the left = 3

The number on the previous tile = 9

The number on this tile = 9 + 3 = 12

Fourth tile:

The number of sides = 3

The number of sides excluding the shared side on the left = 2

The number on the previous tile = 12

The number on this tile = 12 + 2 = 14

Hence, the missing number on the last tile is 14. Have you noticed that we must analyze the shape of the title and its relationship with the number on it in order to solve the problem? In other words, solving problems like this requires analyzing the shape (sides) of the tile and the number on the tile, which improves both *numerical ability* and the *creative side* of the brain. Now that we have found the answers to both problems, we can show them in the next section.

5.3 Answer

Here are the answers to the two problems:

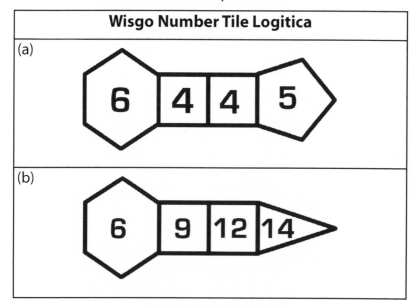

Figure 5.7: Answers

5.4 Summary

We discussed two *Wisgo Number Tile* problems in this chapter. We followed a step-by-step approach to solve both problems. The first problem was straightforward, as the numerical value on each tile was the same as the number of sides each tile had. The second problem was a bit tricky, as we needed to analyze the pattern of numbers on the tiles and relate it to the number of sides of each tile.

In summary, to solve a *Wisgo Tiles* problem, we need to examine the shape of the tile, observe the number indicated on the tile and determine how the two are related. This is what makes this *Logitica so interesting*.

While we have discussed only a few shapes so far, we can also create various types of *Wisgo Number Tile* problems using different

Wisgo Numer Tiles

combinations of shapes. For example, a *Wisgo Number Tile* problem that uses an irregular shape is shown above.

The *author* has created a variety of wisgo problems that are designed to stimulate both the left and right sides of the brain. In general, the *left-side of the brain* is related to the numerical, logical and analytical part of our thinking while the *right-side of the brain* is associated with imagination, art, and creativity. In a *Wisgo Family* problem, there is typically a close relation between the number displayed on a shape (e.g., tiles, grids, etc.), and the geometrical attributes of the shape (e.g., number of sides). The *author* believes that while solving such types of problems, one needs to analyze the *shapes* and *the numbers* related to them. Analyzing shapes stimulates the creative part of the brain (*right-side brain*) and analyzing numerical problems enhances the numerical and logical part of the brain (*left-side brain*). Therefore, these types of problems simultaneously stimulate both sides of the brain and thus assist in the overall development of the brain.

We have provided a variety of problems in the *Exercise Section* so that readers can enjoy learning this innovative *Logitica*.

5.5 Exercise

Find the missing numbers in the problems below:

Level-I

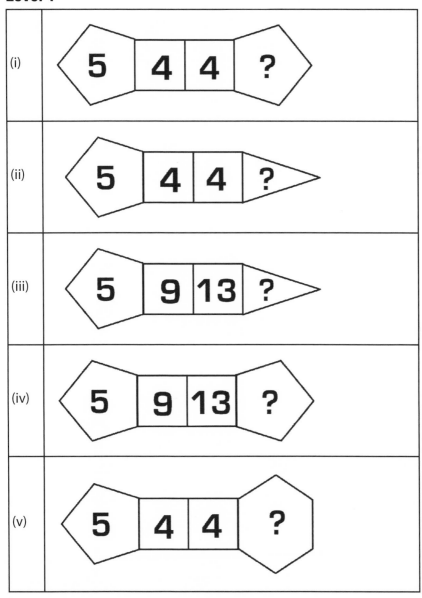

(i)	5 4 4 ?
(ii)	5 4 4 ?
(iii)	5 9 13 ?
(iv)	5 9 13 ?
(v)	5 4 4 ?

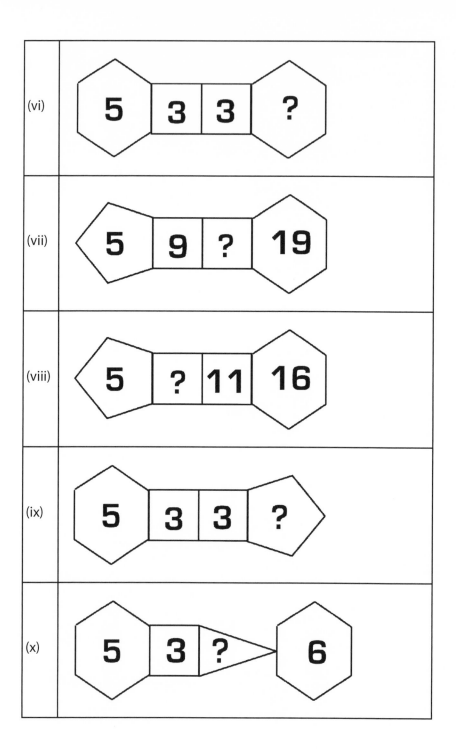

(vi) 5 3 3 ?

(vii) 5 9 ? 19

(viii) 5 ? 11 16

(ix) 5 3 3 ?

(x) 5 3 ? 6

(xi)	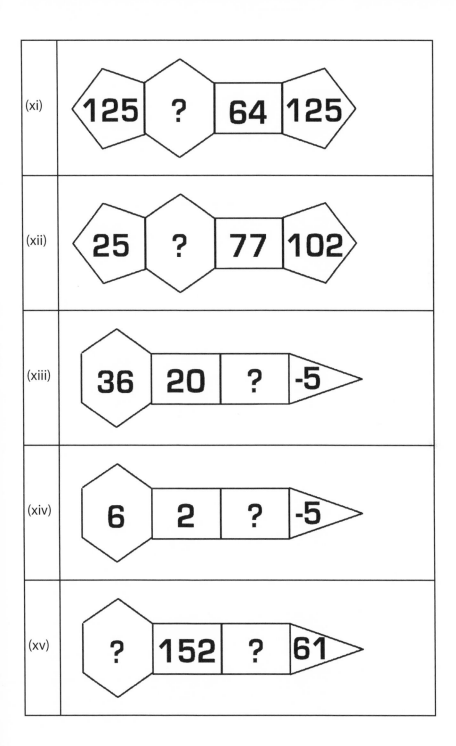
(xii)	
(xiii)	
(xiv)	
(xv)	

(xi) 125 ? 64 125

(xii) 25 ? 77 102

(xiii) 36 20 ? -5

(xiv) 6 2 ? -5

(xv) ? 152 ? 61

Level-II

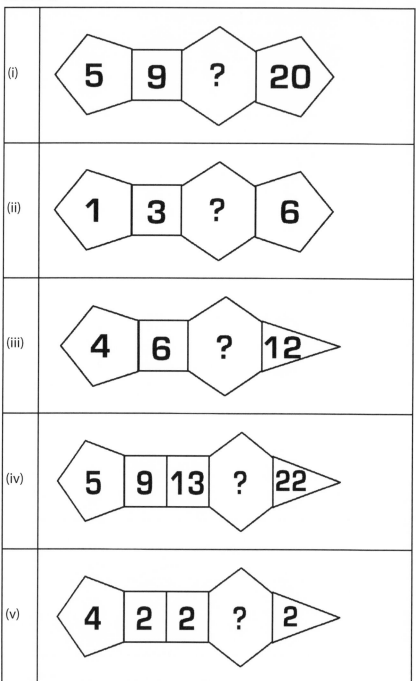

(i) 5 9 ? 20

(ii) 1 3 ? 6

(iii) 4 6 ? 12

(iv) 5 9 13 ? 22

(v) 4 2 2 ? 2

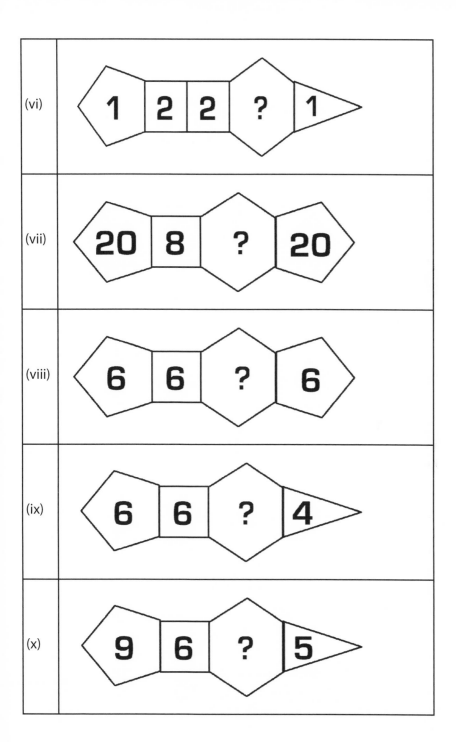

(vi)	1 2 2 ? 1
(vii)	20 8 ? 20
(viii)	6 6 ? 6
(ix)	6 6 ? 4
(x)	9 6 ? 5

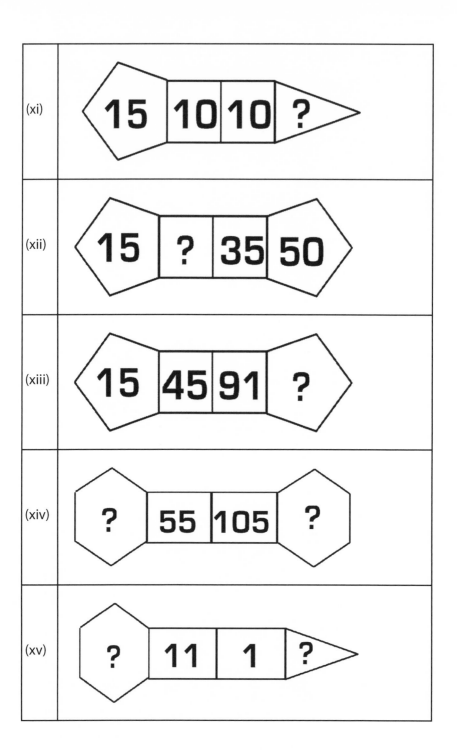

(xi) 15 10 10 ?

(xii) 15 ? 35 50

(xiii) 15 45 91 ?

(xiv) ? 55 105 ?

(xv) ? 11 1 ?

Chapter 6

Number Pyramid

Keywords: Linear Equations, Pascal's Triangle.

6.1 Challenge

Solve the following *Number pyramid* problems:

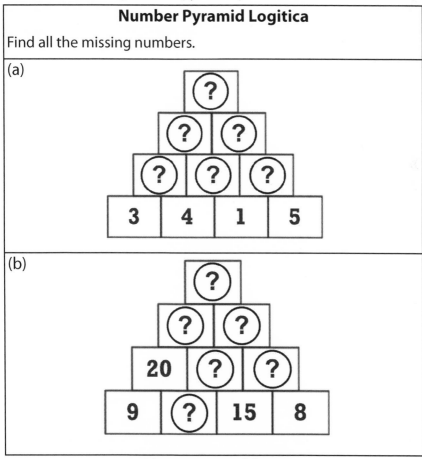

Number Pyramid Logitica
Find all the missing numbers.

(a)

(b)

Figure 6.1: Problems

In this chapter, we will learn how to solve a *Number Pyramid* problem. To start the discussion, let us first define what a *Number Pyramid* is.

♕ **Definition - Number pyramid**

A **Number Pyramid** is a collection of cells arranged in a pyramid. Each cell in the pyramid contains a number, which is the sum of two numbers in the cells directly below it. This rule obviously does not apply to the bottom row, as it has no row below it. It is worth noting that the numbers in a pyramid can be *positive*, *negative* or *fractional*. The aim of a *Number Pyramid* problem is to determine all the missing numbers in the pyramid.

Objective: The aim of a *Number Pyramid* problem is to complete the pyramid. In other words, you need to determine all the missing numbers in the pyramid.

♜ **Note**: For brevity, we will refer to a *Number Pyramid* as a "pyramid" in this chapter.

6.2 Strategy

In this section, we will give an overview of the basic principles behind a *Number Pyramid*. We will also learn how to derive and solve simple equations when solving the pyramid problem (b).

6.2.1 Overview

To find all the missing numbers in a pyramid, we need to determine all the numbers in the bottom row. Once we know all the numbers in the bottom row, we can complete the pyramid by adding the numbers going upwards starting from the bottom row (row-1).

6.2.2 Analysis

Consider an example of a 3-row pyramid as shown in figure-6.2.

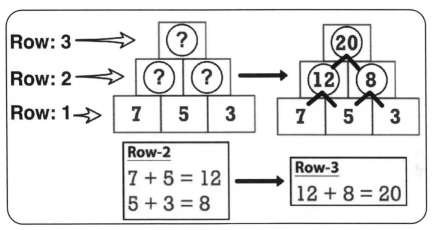

Figure 6.2: A Number Pyramid

In this pyramid, all the numbers in the bottom row are known. A step-by-step approach is provided below to find the missing numbers in the pyramid. These steps are displayed in figure-6.2.

✔ **Step 1**: We first determine the missing numbers in row-2 by adding the numbers from row-1:

$$7 + 5 = 12$$
$$5 + 3 = 8$$

✔ **Step 2**: In this step, we find the missing numbers in row-3 by adding the numbers from row-2:

$$12 + 8 = 20$$

Notice how we add the numbers iteratively upwards to find the missing numbers. Even if you come across a similar pyramid problem with a higher number of rows, you need to follow similar steps to find all the missing numbers. In this chapter, we need to solve two problems. We will analyze each problem while solving it later in the chapter. Let us first define a step-by-step approach on how to solve a pyramid problem.

6.2.3 Solving Steps

We can follow the steps below to solve a pyramid problem:

✔ If we know all the numbers in the bottom row, we can simply add the numbers iteratively in the upward direction to complete the pyramid. We will discuss this approach when solving the pyramid problem (a).

✔ If there is a missing number in the bottom row, we first need to determine that number to solve the pyramid problem. Sometimes we have to derive and solve a simple equation to find the missing number. We will learn more about this approach when solving the pyramid problem (b).

♜ **Note: Advanced Number Pyramid**

However, a pyramid problem can be made a bit more challenging by designing the problem differently. A few examples of such problems can be seen in *Chapter-7* (p. 129). To solve such problems, we need to learn some advanced concepts, which we will discuss in the next chapter.

6.2.4 Solving the Problems

Problem (a)

This is a 4-row pyramid problem. We have seen how to solve a 3-row pyramid earlier in figure-6.2. We will apply the same approach to solve this problem. As shown in figure-6.3, notice how we are calculating the numbers in each cell by adding together the two numbers in the cells directly below it.

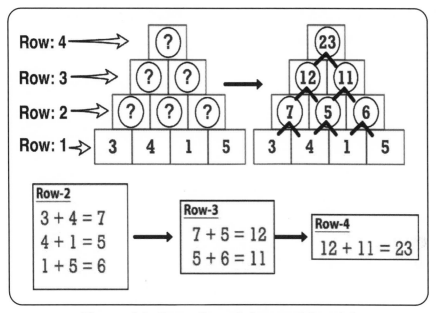

Figure 6.3: Steps for solving problem (a)

A step-by-step approach is provided below to find the missing numbers in the pyramid:

✔ **Step 1**: We first determine the missing numbers in row-2 by adding the numbers from row-1:

$$3 + 4 = 7$$
$$4 + 1 = 5$$
$$1 + 5 = 6$$

✔ **Step 2**: In this step, we find the missing numbers in row-3 by adding the numbers from row-2:

$$7 + 5 = 12$$
$$5 + 6 = 11$$

✔ **Step 3**: Since we are solving a 4-row pyramid, the number in the topmost cell of the pyramid is obtained by adding the numbers from row-3:

$$12 + 11 = 23$$

Now that we have found all the missing numbers, we can show the completed pyramid in the *Answer Section* of the chapter. The purpose of this example is to explain the concept behind the calculations needed to find the missing numbers in a pyramid. A basic understanding of this concept is essential, as it will help you to solve more complicated pyramid problems.

Problem (b):

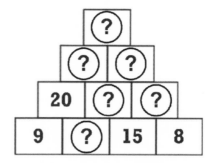

Figure 6.4: Problem (b)

We can follow a step-by-step approach to find the missing numbers as described below. We have also summarized these steps later in figure-6.5:

✔ **Step 1:** We should first try to find cells for which we can calculate the numbers quickly by using the known numbers in the pyramid. In this problem, we can see that number 15 and 8 in the bottom row are next to each other. Hence, we can use these numbers to find the number in the cell above them:

$$15 + 8 = 23$$

However, we cannot calculate any more numbers in row-2 as we have a missing number in the bottom row. In the subsequent steps, we will learn how to determine the remaining missing numbers in the pyramid.

Adding numbers in the pyramid

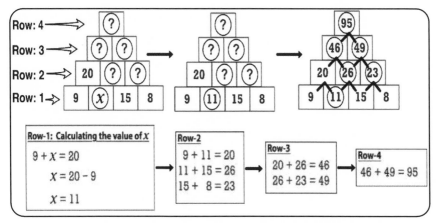

Figure 6.5: Steps for solving problem (b)

✔ **Step 2**: Let us label the missing number in the bottom row as x. Since we know that the value of any cell in the pyramid is the sum of the two numbers directly below it, we can use the following equation to determine the value of x:

$$9 + x = 20$$

$$x = 20 - 9 = 11$$

✔ **Step 3**: Now that we know all the numbers in the bottom row of the pyramid, we can easily calculate the numbers in row-2:

$$11 + 15 = 26$$

✔ **Step 4**: The numbers in row-3 can now be calculated by adding the numbers from row-2:

$$20 + 26 = 46$$
$$26 + 23 = 49$$

✔ **Step 5**: We can now calculate the number at the top of the pyramid by adding the numbers from row-3:

$$46 + 49 = 95.$$

It is easy to see that once we have all the numbers in the bottom row, the numbers in the upper rows can easily be calculated by adding the appropriate numbers together iteratively upwards. Now that we have determined all the missing numbers in the pyramid, we can show the *completed pyramid* in the next section.

6.3 Answer

The answers to the problems are shown below:

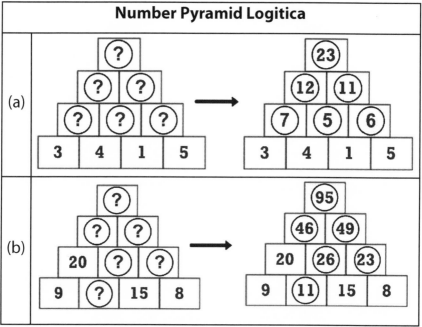

Figure 6.6: Answers

6.4 Summary

In this chapter, we discussed how to solve the pyramid problems. We started the chapter with a simple pyramid in problem (a). We learned how to use the numbers in the lower to obtain the numbers in the upper row. However, to solve the problem (b), we had to write an equation and solve it to determine the missing number at the bottom row. A pyramid problem can be made more challenging by designing the problem differently, which we will discuss in the next chapter.

The contents of this chapter provide an excellent foundation for solving simple pyramid problems. Readers are advised to solve the problems in the *Exercise Section* to get a better understanding of the concepts involved.

6.5 Exercise

Find the missing numbers in the problems below:

Level-I

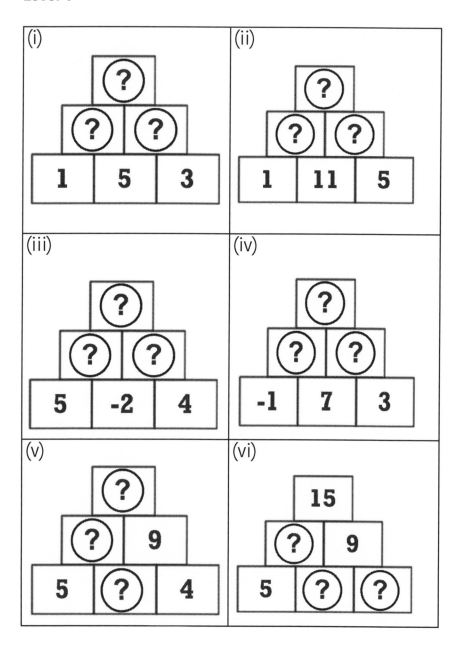

(i)
?
? ?
1 5 3

(ii)
?
? ?
1 11 5

(iii)
?
? ?
5 -2 4

(iv)
?
? ?
-1 7 3

(v)
?
? 9
5 ? 4

(vi)
15
? 9
5 ? ?

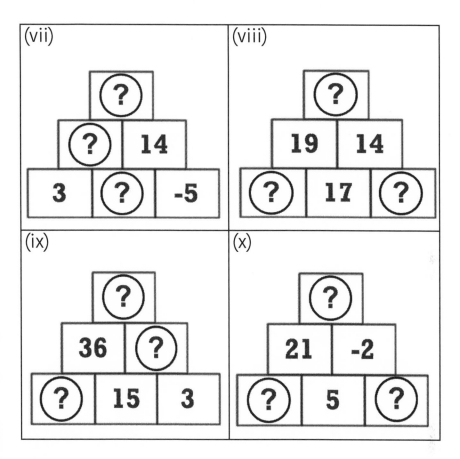

Level-II

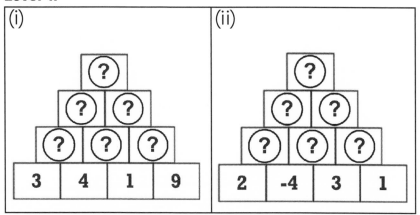

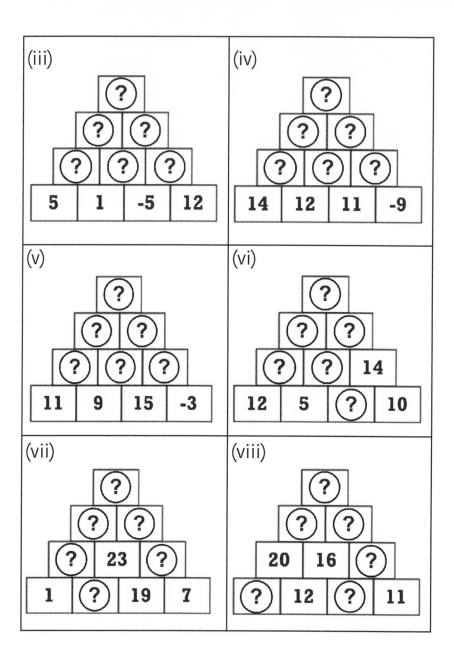

(iii)

?

? ?

? ? ?

| 5 | 1 | -5 | 12 |

(iv)

?

? ?

? ? ?

| 14 | 12 | 11 | -9 |

(v)

?

? ?

? ? ?

| 11 | 9 | 15 | -3 |

(vi)

?

? ?

? ? 14

| 12 | 5 | ? | 10 |

(vii)

?

? ?

? 23 ?

| 1 | ? | 19 | 7 |

(viii)

?

? ?

20 16 ?

| ? | 12 | ? | 11 |

(ix)

(x)

(xi)

(xii)

(xiii)

(xiv)

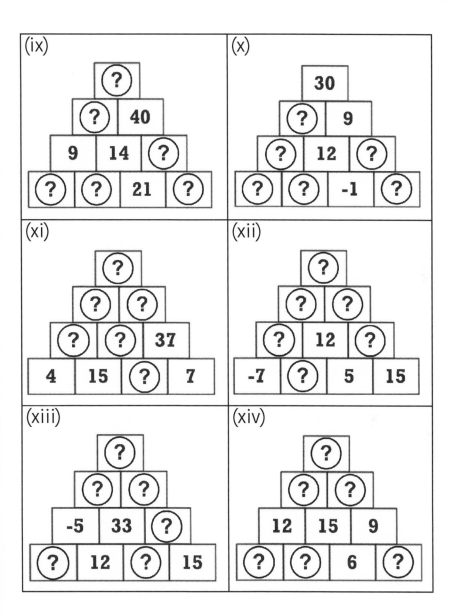

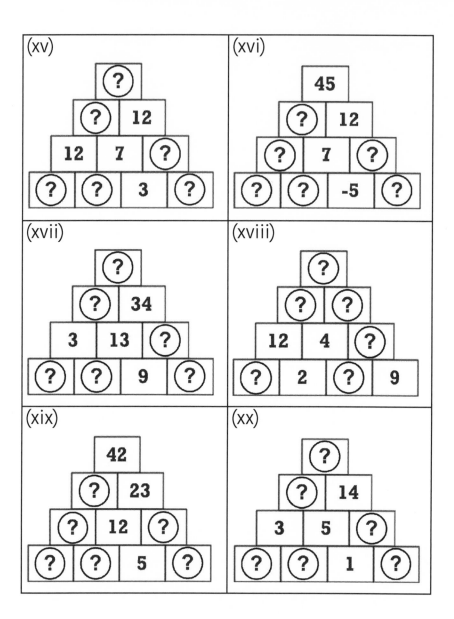

(xv)

(xvi)

(xvii)

(xviii)

(xix)

(xx)

Chapter 7

Advanced Number Pyramid

Keywords: Linear Equations, Pascal's Triangle.

7.1 Challenge

Solve the following *Advanced Number pyramid* problems:

Advanced Number Pyramid Logitica
Find all the missing numbers.

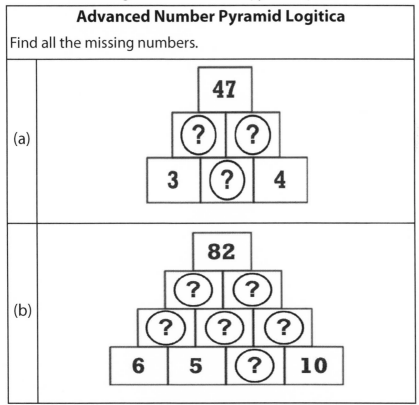

Figure 7.1: Problems

In this chapter, we will learn how to solve an *Advanced Number Pyramid* problem. To start the discussion, let us first define what we mean by an *Advanced Number Pyramid* problem.

♛ Definition - Advanced Number pyramid

We have discussed *Number Pyramid* in **Chapter-6** (p. 115). So if you are not familiar with how to solve a simple pyramid problem, please refer to the previous chapter for a basic understanding of the topic.

Similar to what is shown in figure-7.1, these types of pyramid problems cannot be solved by simply adding the numbers or writing simple equations as we did in the previous chapter. To solve such types of problems, we need to learn some more concepts such as *Pascal's triangle* and *Coefficient Rules*. That is why we refer to such problems as *Advanced Number Pyramid problems.* We will discuss these concepts in this chapter and learn how to apply them in solving such types of pyramid problems.

Objective: The aim of a pyramid problem is to complete the pyramid. In other words, you need to determine all the missing numbers in the pyramid.

7.2 Strategy

In this section, we will give an overview of an *Advance Number Pyramid* problem followed by a thorough analysis of the topic. We will later outline steps describing how to solve such problems. We will subsequently solve the pyramid problems from this chapter using the concepts we discussed in the *Analysis Section*.

7.2.1 Overview

The *Number Pyramid* problems in this chapter are slightly advanced than what we have solved in the previous chapter. In such type of problems, one of the numbers in the bottom row is missing along with all the numbers in the middle rows. To solve a pyramid problem of such complexity, we need to write an equation involving the numbers in the cells from two different rows. We will see later that *Pascal's triangle* is used in determining the coefficients of these equations.

7.2.2 Analysis

To solve problems in this chapter, we will need to learn some advanced concepts like *Pascal's triangle* and *Coefficient rules*. Let us first discuss *Pascal's triangle*.

7.2.1.1 Pascal's triangle

Pascal's triangle (named after Blaise Pascal, a famous French mathematician) is a triangular arrangement of numbers.

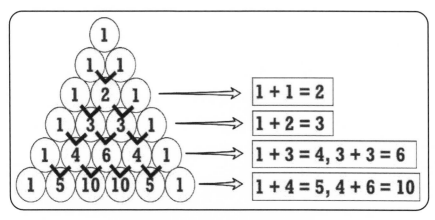

Figure 7.2: Pascal's triangle

To build a *Pascal's triangle*, we start with number 1 in the top cell. We then calculate the numbers for the interior cells by adding the two numbers directly above each cell as shown in figure-7.2. The numbers in the exterior cells on both sides of the *triangle* are always 1. As you can see in the figure below, addition starts from the second row because you need at least two numbers to add together.

♜ **Note**: *Pascal's triangle* gives the coefficients in the expansion of a *binomial expression*. A discussion on *binomial expression* is beyond the scope of this book. Refer to any standard textbook for a brief introduction on the topic.

In a *Number Pyramid*, there is a relation between the numbers in the lower and upper rows. This relation depends on how far the two rows are separated from each other. We will write an equation to represent this relation *mathematically* using the *Coefficient Rules* that will be discussed in the next section.

7.2.2.2 Coefficient Rules

We will use the following *Coefficient Rules* to solve pyramid problems:

i.   **(1, 1) Coefficient Rule**

ii.   **(1, 2, 1) Coefficient Rule**

iii. **(1, 3, 3, 1) Coefficient Rule**

iv.   **Higher-coefficient Rules**

Each of these rules is derived from Pascal's triangle. In the discussion below, we will call Pascal's triangle, *for brevity*, "triangle."

Referring to Pascal's triangle in figure-7.2:

- (1, 1) *Coefficient Rule* is derived from the two numbers (1, 1) in the second row from the top of the triangle.

- (1, 2, 1) *Coefficient Rule* is derived from the three numbers (1, 2, 1) in the third row from the top of the triangle.

- (1, 3, 3, 1) *Coefficient Rule* is derived from the four numbers (1, 3, 3, 1) in the fourth row from the top of the triangle.

- Similarly, we can define *Higher-Coefficient Rules* using the numbers from the appropriate rows of the triangle.

Pascal's triangle

To explain the *Coefficient Rules*, we will use a 4-row pyramid. A 4-row pyramid contains 10 numerical cells, whose values can be represented by ten variables: $x_1, x_2, x_3, x_4, x_5, x_6, x_7, x_8, x_9$, and x_{10}. Hence, these rules, as they apply to the numbers in this pyramid, will also apply to these ten variables. You will see further down that we have filled in the relevant cells of the pyramid with *black* to indicate which cells have been selected in

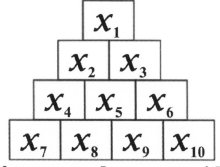

4-row number pyramid

the discussion of the corresponding rule. In the following section, we will discuss each of the rules individually using some examples:

i. (1, 1) Coefficient Rule: We use this rule when formulating an equation involving a variable in the upper row and two variables directly under it in the row beneath as shown in figure-7.3 and figure-7.4. Notice how (1, 1) is used as the coefficients of the two variables from the row beneath.

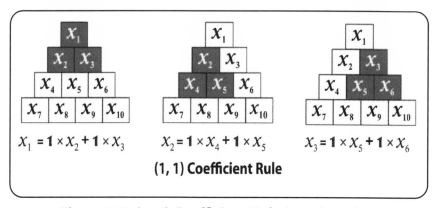

Figure 7.3: (1, 1) Coefficient Rule (continued...)

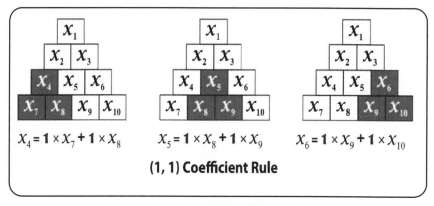

Figure 7.4: (1, 1) Coefficient Rule

Let us review the following equation defined for the first pyramid in figure-7.3:

$$X_1 = (1 \times X_2) + (1 \times X_3) = X_2 + X_3$$

The (1, 1) *Coefficient Rule* refers to the coefficients of X_2 and X_3 in this equation. Similarly, we can observe the other uses of this rule for a 4-row pyramid as shown in figure-7.3 and figure-7.4. This rule is easy to understand, and we have given a few examples of this rule in figure-7.5 and figure-7.6. In a 4-row pyramid, there are six ways in which we can select the cells that this rule applies to.

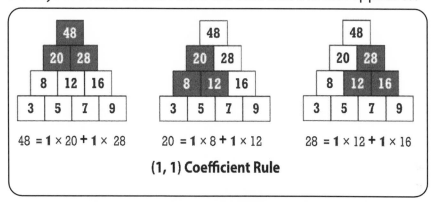

Figure 7.5: Examples of (1, 1) Coefficient Rule (continued...)

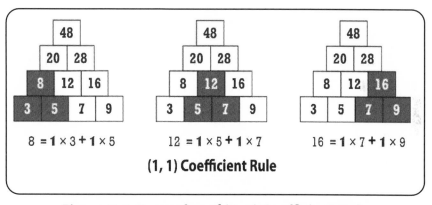

Figure 7.6: Examples of (1, 1) Coefficient Rule

ii. (1, 2, 1) Coefficient Rule: This rule defines the relationship between variables in cells that are separated by a row in between. We use this rule to write an equation that relates a variable in the upper row to the three variables in the lower row as shown in figure-7.7. Here (1, 2, 1) represents the coefficients of the three variables in the lower row as described below:

(a) The first pyramid in figure-7.7: In this pyramid, (1, 2, 1) represents the respective coefficients of variables $X_4, X_5,$ and X_6 in the equation involving X_1 and these three variables as shown below:

$$X_1 = (1 \times X_4) + (2 \times X_5) + (1 \times X_6) = X_4 + 2X_5 + X_6$$

(b) The second pyramid in figure-7.7: In this pyramid, (1, 2, 1) represents the respective coefficients of variables $X_7, X_8,$ and X_9 in the equation involving X_2 and these three variables as shown below:

$$X_2 = (1 \times X_7) + (2 \times X_8) + (1 \times X_9) = X_7 + 2X_8 + X_9$$

(c) The third pyramid in figure-7.7: In this pyramid, (1, 2, 1) represents the respective coefficients of variables $X_8, X_9,$ and X_{10} in the equation involving X_3 and these three variables as shown below:

$$X_3 = (1 \times X_8) + (2 \times X_9) + (1 \times X_{10}) = X_8 + 2X_9 + X_{10}$$

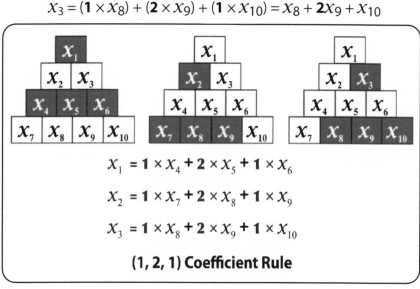

$$X_1 = 1 \times X_4 + 2 \times X_5 + 1 \times X_6$$
$$X_2 = 1 \times X_7 + 2 \times X_8 + 1 \times X_9$$
$$X_3 = 1 \times X_8 + 2 \times X_9 + 1 \times X_{10}$$

(1, 2, 1) Coefficient Rule

Figure 7.7: (1, 2, 1) Coefficient Rule

Examples of the (1, 2, 1) *Coefficient Rule* are shown in figure-7.8. As shown in the figure, when using the (1, 2, 1) rule, there are three ways to select cells in a 4-row pyramid. In each case, the two rows are separated by one row in between.

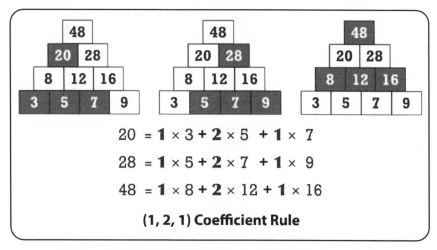

$$20 = 1 \times 3 + 2 \times 5 + 1 \times 7$$
$$28 = 1 \times 5 + 2 \times 7 + 1 \times 9$$
$$48 = 1 \times 8 + 2 \times 12 + 1 \times 16$$

(1, 2, 1) Coefficient Rule

Figure 7.8: Examples of (1, 2, 1) Coefficient Rule

iii. (1, 3, 3, 1) Coefficient Rule: This rule defines the relationship between variables in cells that are separated by two rows in between. We can write an equation relating a variable in the upper row to the four variables in the lower row using this rule. Here (1, 3, 3, 1) are the coefficients of four variables in the lower row. For example, as shown in figure-7.9, (1, 3, 3, 1) are the respective coefficients of the four variables x_7, x_8, x_9, and x_{10} in the equation involving x_1 and these four variables:

$$x_1 = (1 \times x_7) + (3 \times x_8) + (3 \times x_9) + (1 \times x_{10})$$
$$= x_7 + 3x_8 + 3x_9 + x_{10}$$

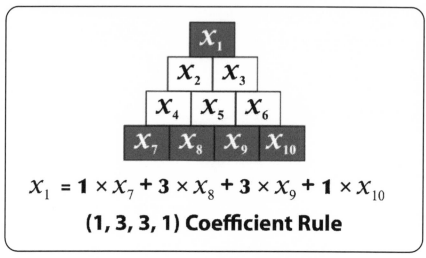

Figure 7.9: (1, 3, 1) Coefficient Rule

An example of (1, 3, 3, 1) *Coefficient Rule* is shown in figure-7.10. It shows that when using this rule, there is only one way to select cells in a 4-row pyramid.

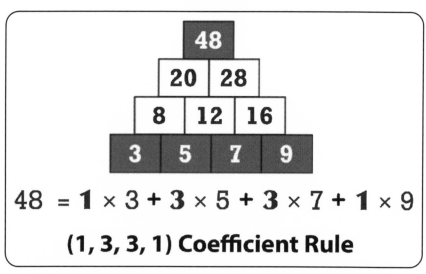

Figure 7.10: An example of (1, 3, 3, 1) Coefficient Rule

Referring to figure-7.9 and figure-7.10, we can assign the following numerical values to $x_1, x_7, x_8, x_9,$ and x_{10}:

$$x_1 = 48, x_7 = 3, x_8 = 5, x_9 = 7, x_{10} = 9$$

$$\begin{aligned} x_1 &= x_7 + 3x_8 + 3x_9 + x_{10} \\ &= 3 + 3 \times 5 + 3 \times 7 + 9 \\ &= 3 + 15 + 21 + 9 \\ &= 48 \end{aligned}$$

Did you notice that you can now calculate the number at the top by using the numbers from the bottom row in the formula: $x_1 = x_7 + 3x_8 + 3x_9 + x_{10}$? Hence, we can calculate the value of x1 in a single step using the (1, 3, 3, 1) *Coefficient Rule*. In summary, now that you know how to use these rules, you no longer need to calculate the numbers iteratively upwards to determine the number at the top. As you will see later, we will use the (1, 3, 3, 1) *Coefficient Rule* to solve the problem (b).

iv. Higher-Coefficient Rules: If you come across a pyramid with more than 4 rows, you will need *Higher-Coefficient Rules* to solve such problems. These rules can be derived using *Pascal's triangle*. We have provided a derivation of these rules for pyramids up to 5 rows in **Appendix-E** (p. 171). We have also summarized the key points of the *Coefficient Rules* in the **Summary** (p.122) section of this chapter.

🐦 Rule of thumb: Using the Coefficient Rules

Pascal's triangle gives the coefficients that are used in the equation relating numbers from two different rows. Therefore, it is essential to understand the basic concept of *Pascal's triangle* and how it is used in solving pyramid problems. We can summarize the *Coefficient Rules* as shown below:

When two rows are:

- ✔ **stacked together with *no rows* in between:** Refer to the (1, 1) *Coefficient Rule*,
- ✔ **separated by *one row* in between**: Refer to the (1, 2, 1) *Coefficient Rule*,
- ✔ **separated by *two rows* in between**: Refer to the (1, 3, 3, 1) *Coefficient Rule*,
- ✔ **separated by *more rows* in between**: Refer to the *Higher-Coefficient Rule*.

7.2.3 Solving Steps

We can follow the steps below to solve the *Advanced Number Pyramid* problems:

- ✔ Identify two rows that will be used in solving the pyramid, and count the number of rows between these two rows. We use this number count to select the appropriate *Coefficient Rules* for solving the problem.
- ✔ Write the relevant equations using the numbers from different rows and *Coefficient Rules*. Once we have the required number of equations, we can solve them to find the missing numbers.
- ✔ Note that once we have all the numbers in the bottom row of the pyramid, all the remaining numbers can be calculated by simply adding the numbers in the upward direction.

7.2.4 Solving the Problems

Let us try solving problem (a) using the steps laid out in the previous section.

Problem (a)

The pyramid in problem (a) contains 3 rows. We know the number at the topmost cell of the pyramid. Also, one of the numbers in the bottom number is missing.

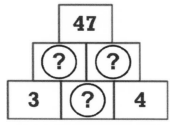

Figure 7.11: Problem (a)

Let us follow a step-by-step approach to solve this problem:

✔ **Step 1**: The number in the topmost cell is 47 and one of the numbers in the bottom row is missing. Let us assume this missing number is x. So the values of the cells in the bottom row are 3, x, and 4.

✔ **Step 2**: We are going to use the topmost row and bottom row to solve this *Number Pyramid* problem. Since these two rows are separated by 1 row in between, we will use (1, 2, 1) *Coefficient Rule* to find the value of x as shown below:

$$1 \times 3 + 2x + 1 \times 4 = 47$$

$$3 + 2x + 4 = 47$$

$$2x + 7 = 47$$

$$2x = 47 - 7 = 40$$

$$x = \frac{40}{2} = 20$$

✔ **Step 3**: Now that we know all the numbers in the bottom row, we can find the remaining numbers as described below:

Row-2:

$$3 + 20 = 23$$

$$20 + 4 = 24$$

✔ **Step 4**: This is the verification step. If our calculations are correct so far, then the number at the top of the pyramid should equal the sum of the two numbers directly below it. Since the numbers in row-2 are 23 and 24, the number in the row-3 is $23 + 24 = 47$, which equals the number in the topmost cell. This verifies that our calculated numbers are correct.

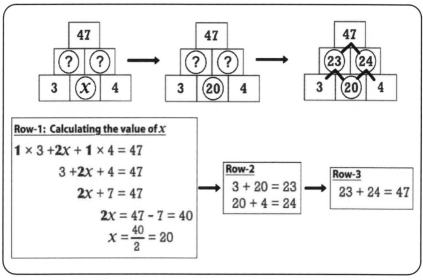

Figure 7.12: Steps for solving problem(a)

Problem (b)

We will now solve a 4-row pyramid in problem (b). We will have to use the number in the topmost cell to find the missing values in the pyramid, along with the numbers we do have in row-1.

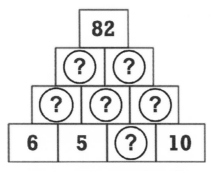

Figure 7.13: Problem (b)

We can follow a step-by-step approach to solve this problem as described below:

✔ **Step 1**: Let us first try to identify the cells in which we can quickly determine the missing numbers by using the known numbers in the pyramid. As shown in the figure-7.13, the numbers 6 and 5 are next to each other in the bottom row. These numbers can be immediately added together to calculate the number for the cell directly above them as shown below:

$$6 + 5 = 11$$

Adding numbers in the pyramid

We must now find the missing number in the bottom row before we can proceed any further, which we will do in the next step.

✔ **Step 2**: In this pyramid, we know the number in the topmost cell and all the numbers in the bottom row except for one. Since the top and bottom rows are separated by two rows, we need to use the (1, 3, 3, 1) *Coefficient Rule* to write an equation to determine the missing number in the bottom row. Let us call the missing number in the bottom row, x. So the values of cells in the bottom row are 6, 5, x, and 10 in the bottom row.

We can now use this rule to find the value of x as shown below:

$$1 \times 6 + 3 \times 5 + 3x + 1 \times 10 = 82$$

$$6 + 15 + 3x + 10 = 82$$

$$31 + 3x = 82$$

$$3x = 82 - 31 = 51$$

$$x = \frac{51}{3} = 17$$

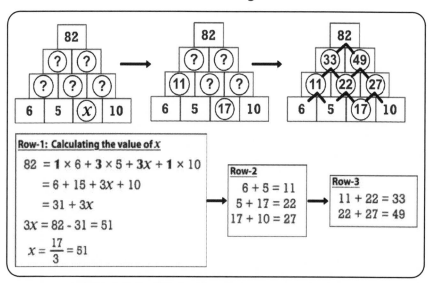

Figure 7.14: Steps for solving problem(b)

✔ **Step 3**: Now we can easily determine the missing numbers that remain in the pyramid by adding the appropriate numbers together going upwards as shown below:

Row-2:

$$5 + 17 = 22$$

$$17 + 10 = 27$$

Row-3:

$$11 + 22 = 33$$
$$22 + 27 = 49$$

Row-4: This is the verification step. If our calculations are correct so far, then the number at the top of the pyramid should equal the sum of the two numbers directly below it. Since the numbers in row-3 are 33 and 49, the number in the row-4 is 33 + 49 = 82, which equals the number in the topmost cell. This verifies that our calculated numbers are correct.

Now that we have solved both the problems, we can display the completed pyramids in the next section.

7.3 Answer

The answers to the problems are shown below:

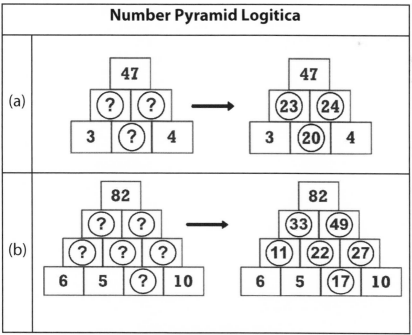

Figure 7.15: Answers

♖ Note on using Coefficient Rules:

We have used the (1, 3, 3, 1) *Coefficient Rule* to solve the problem in this example, but whenever needed, we can also use other *Coefficient Rules*. For example, notice how we have used the (1, 2, 1) *Coefficient Rule* in each of the pyramids below:

Example-1: For the pyramid on the right side, the two rows containing numbers (37) and (11, 7, 12) are separated by one row in between, the numbers in these two rows are related by (1, 2, 1) *Coefficient Rules* as shown below:

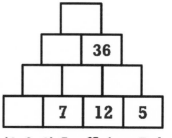

(1, 2, 1) Coefficient Rule

$1 \times 11 + 2 \times 7 + 1 \times 12 = 11 + 14 + 12 = 37$

Example-2: For the pyramid on the right side, the two rows containing numbers (36) and (7, 12, 5) are separated by one row in between, the numbers in these two rows are related by (1, 2, 1) *Coefficient Rules* as shown below:

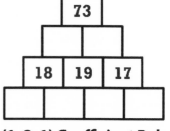

(1, 2, 1) Coefficient Rule

$1 \times 7 + 2 \times 12 + 1 \times 5 = 7 + 24 + 5 = 36$

Example-3: For the pyramid on the right side, the two rows containing numbers (73) and (18, 19, 17) are separated by one row in between, the numbers in these two rows are related by (1, 2, 1) *Coefficient Rules* as shown below:

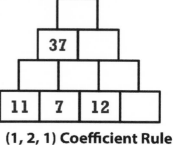

(1, 2, 1) Coefficient Rule

$1 \times 18 + 2 \times 19 + 1 \times 17 = 18 + 38 + 17 = 73$

Have you noticed that using the appropriate *Coefficient Rule*, we can calculate the number in a cell directly without needing to add the numbers iteratively in the upward direction? That is the benefit of using *Coefficient Rules* when solving an *Advanced Number Pyramid problem.*

7.4 Summary

In this chapter, we learned the various concepts required for solving an *Advanced Number Pyramid* problem. We had two problems to solve in this chapter, for which we needed a good understanding of *Pascal's triangle* and the various *Coefficient Rules*. In this chapter, we have defined three *Coefficient Rules* using *Pascal's triangle*. We have used a 4-row pyramid to explain how to use the *Coefficient Rules* for writing equations that can be solved for finding the missing numbers in the pyramid. To solve a problem involving a pyramid with a higher number of rows, you need to derive *Higher-Coefficient Rules*. In the discussion below, we will call Pascal's triangle, for brevity, "triangle." As shown below, we have extended the list of *Coefficient Rules* using Pascal's triangle.

When the selected two rows in a pyramid are:

✔ **stacked together with no rows in between:** Use the (1, 1) *Coefficient Rule.* This rule is derived from the two numbers (1, 1) in the second row from the top of the triangle and applies to a pyramid with a *minimum of 2 rows.*

✔ **separated by one row in between:** Use the (1, 2, 1) *Coefficient Rule.* This rule is derived from the three numbers (1, 2, 1) in the third row from the top of the triangle and applies to a pyramid with a *minimum of 3 rows.*

✔ **separated by two rows in between:** Use the (1, 3, 3, 1) *Coefficient Rule*. This rule is derived from the four numbers (1, 3, 3, 1) in the fourth row from the top of the triangle and applies to a pyramid with a *minimum of 4 rows.*

✔ **separated by three rows in between:** Use the (1, 4, 6, 4, 1) *Coefficient Rule*. This rule is derived from the five numbers (1, 4, 6, 4, 1) in the fifth row from the top of the triangle and applies to a pyramid with a *minimum of 5 rows.*

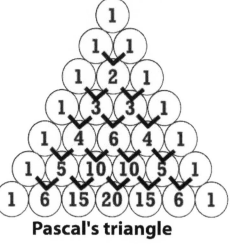

Pascal's triangle

✔ **separated by four rows in between:** Use the (1, 5, 10, 10, 5, 1) *Coefficient Rule*. This rule is derived from the six numbers (1, 5, 10, 10, 5, 1) in the sixth row from the top of the triangle and applies to a pyramid with a *minimum of 6 rows.*

✔ **separated by five rows in between:** Use the (1, 6, 15, 20, 15, 6,1) *Coefficient Rule*. This rule is derived from the seven numbers (1, 6, 15, 20, 15, 6,1) in the seventh row from the top of the triangle and applies to a pyramid with a *minimum of 7 rows.*

✔ **separated by more rows in between:** Refer to *Pascal's triangle* for the appropriate *Coefficient Rule* to use. Based on the number of rows in the pyramid, we need to select the numbers from the appropriate row of the triangle. These numbers will be the coefficients in the corresponding *Coefficient Rule*.

The contents of this chapter provide an excellent foundation for solving various types of pyramid problems. Readers are advised to solve the problems in the *Exercise Section* to get a better understanding of the concepts involved.

7.5 Exercise

Find the missing numbers in the problems below:

Level-I

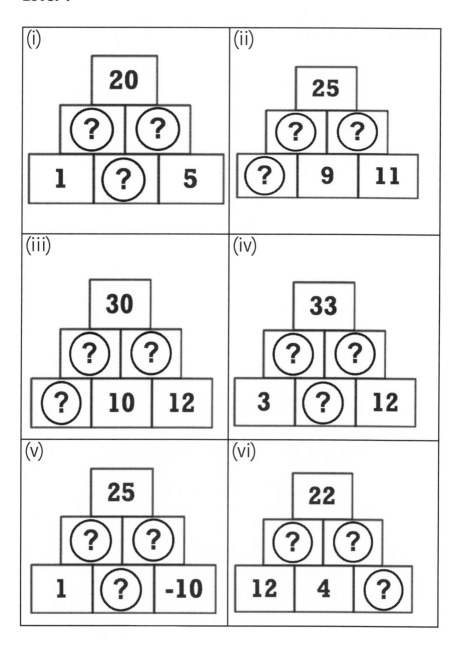

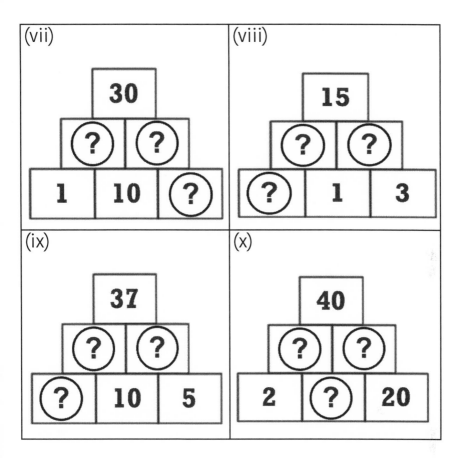

(vii)

30

? ?

1 | 10 | ?

(viii)

15

? ?

? | 1 | 3

(ix)

37

? ?

? | 10 | 5

(x)

40

? ?

2 | ? | 20

Level-II

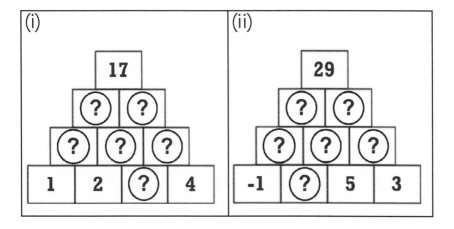

(i)

17

? ?

? ? ?

1 | 2 | ? | 4

(ii)

29

? ?

? ? ?

-1 | ? | 5 | 3

(iii)

(iv)

(v)

(vi)

(vii)

(viii)

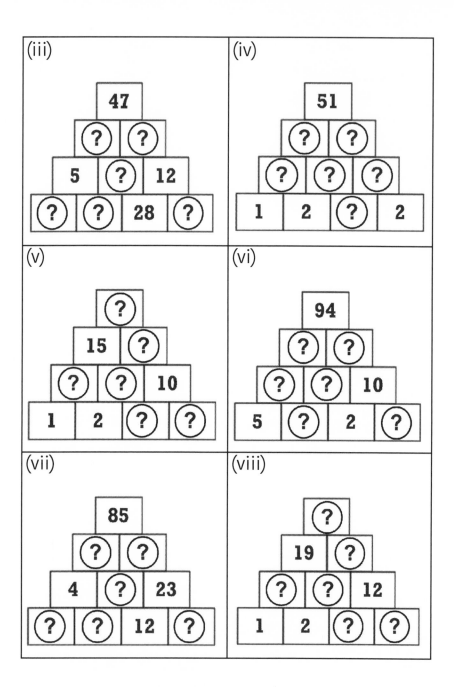

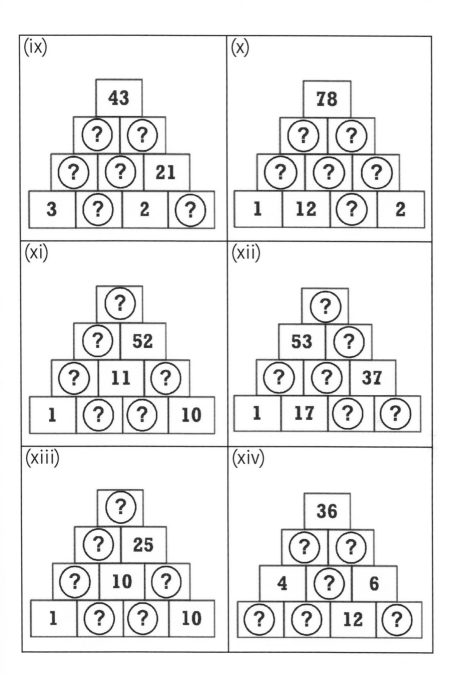

(ix)

(x)

(xi)

(xii)

(xiii)

(xiv)

153

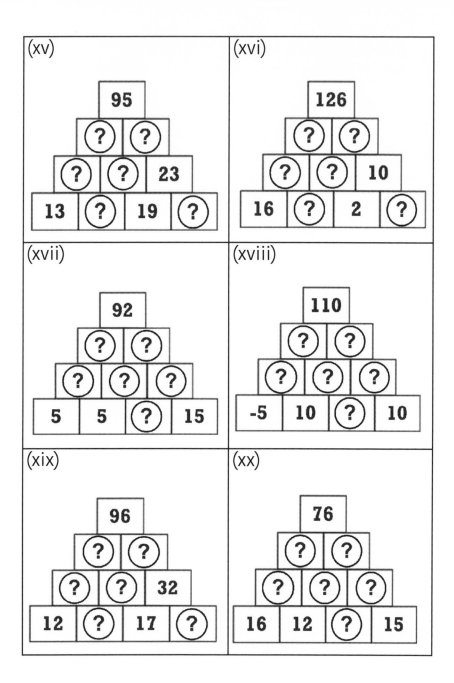

(xv)

95

? ?

? ? 23

13 ? 19 ?

(xvi)

126

? ?

? ? 10

16 ? 2 ?

(xvii)

92

? ?

? ? ?

5 5 ? 15

(xviii)

110

? ?

? ? ?

-5 10 ? 10

(xix)

96

? ?

? ? 32

12 ? 17 ?

(xx)

76

? ?

? ? ?

16 12 ? 15

154

Section-II Appendix

In this section, we have included a few mathematical concepts that are essential for solving problems in this book.

The concepts in this section are presented in a simpler form, focusing mostly on their use for learning and solving problems in this book. For more *rigorous* and *advanced* content on the topics, please refer to the relevant books or literature.

(A) Sequential dependency among Number Boxes

Some books or aptitude tests might have *Number Box* problems that require *sequential dependency* to solve them. This can be better explained with an example below:

Figure-A1: An example of sequential dependency

In figure-A1, we have three boxes. We can use the steps described below to determine the middle number. In the steps below, we will use notations (a_1, b_1), (a_2, b_2), and (a_3, b_3) to represent the numbers on the left and the right sides of the three boxes:

✔ Add the number on the left to the number on the right for each box. The result of this step will be $(a_1 + b_1)$, $(a_2 + b_2)$, and $(a_3 + b_3)$ for the three boxes respectively.

✔ Add 1, 2, and 3 sequentially as per the position of boxes in the problem to the result of the previous step to determine the missing numbers. In other words, the middle numbers in the first, second and third boxes will be $(a_1 + b_1 + 1)$, $(a_2 + b_2 + 2)$, and $(a_3 + b_3 + 3)$ as shown below:

The first box:

$a_1 = 4, b_1 = 1$

$a_1 + b_1 = 4 + 1 = 5$

$a_1 + b_1 + 1 = 5 + 1 = 6$. This equals the middle number 6.

The second box:

$a_2 = 5, b_2 = 1$

$a_2 + b_2 = 5 + 1 = 6$

$a_2 + b_2 + 2 = 6 + 2 = 8$. This equals the middle number 8.

The third box:

$a_3 = 7, b_3 = 2$

$a_3 + b_3 = 7 + 2 = 9$

$a_3 + b_3 + 3 = 9 + 3 = 12$.

This is the answer we get by assuming a sequential dependency in solving the problem. As you will see later, this is not the right way to design a *Number Box problem*.

Notice how we have used numbers 1, 2, and 3 in each of the boxes sequentially to find the answer. This is an example of a sequential dependency among the boxes. Here we have used the natural numbers (1, 2, 3, ...) sequentially to be added to the sum of the numbers on the left and right. In a different problem, another variation of the sequential numbers (e.g., 0, 1, 3, or -1, -2, -3, etc.) can be added to the values that result from applying a different combination of arithmetic operations on the two numbers in the box. While it is possible that some books and aptitude tests might have answers that need to be calculated in this manner, the author believes that all the logic required to find the missing number should be encoded in the individual box, meaning that each box should contain its logic within. Hence, the problems in this book do not assume any form of sequential dependency among the boxes. However, readers are advised to follow the *guidelines and norms* of the books or aptitude tests to solve such problems in those books or tests and prepare accordingly.

This brings us to the question: can the problem in figure-A1 be solved without using a *sequential dependency*? In cases where a problem is designed by assuming a sequential dependency, finding the solution using other ways might not even be possible. However, in this case, we can find an answer as explained below:

✔ We will use notations (a, b) to represent numbers on the left and right sides of the box. To be more specific, we will use notations (a_1, b_1), (a_2, b_2), and (a_3, b_3) to represent the numbers on the left and right sides for the three boxes respectively.

✔ Find the difference between two numbers by subtracting the number on the right from the number on the left. This can be expressed as $(a - b)$.

✔ Double the result from the previous step, which can be formulated as $2(a - b)$. This formula can now be used to determine the middle number in each box as shown below:

The first box:

$a_1 = 4, b_1 = 1$

$a_1 - b_1 = 4 - 1 = 3$

$2(a_1 - b_1) = 2 \times 3 = 6$. This equals the middle number 6.

The second box:

$a_2 = 5, b_2 = 1$

$$a_2 - b_2 = 5 - 1 = 4$$

$2(a_2 - b_2) = 2 \times 4 = 8$. This equals the middle number 8.

The third box:

$$a_3 = 7, b_3 = 2$$

$$a_3 - b_3 = 7 - 2 = 5$$

$2(a_3 - b_3) = 2 \times 5 = 10$. This is the answer to the missing number in the third box.

The answer differs from what we found earlier where we assumed a sequential dependency among the boxes. Did you notice that we solved this problem without assuming any sequential dependency at all? These boxes were merely using the same logic to determine the middle numbers without using any sequential dependency, which is the right way to design such problems.

While it is possible that some books and aptitude tests might have answers that need to be calculated using the sequential dependency, the author believes that all the logic required to find the missing number should be encoded in the individual box, meaning that each box should contain its logic within. Hence, the problems in this book do not assume any form of sequential dependency among the boxes. However, readers are advised to follow the *guidelines and norms* of the books or aptitude tests to solve such problems in those books or tests and prepare accordingly.

(B) Sequential dependency among Number Crosses

Some books or aptitude tests may contain *Number Cross* problems in which *sequential dependency* is required to solve them. This can be better explained with an example below:

Figure-B1: An example of sequential dependency

In figure-B1, we have three crosses, and we need to find the missing number in the third cross. We can use the following steps to determine the missing number:

- Add up all four numbers in the cross. Let us call the totals for each cross c_1, c_2, and c_3 respectively.

- The number in the center is now determined by adding natural numbers (1, 2, and 3) to c_1, c_2, and c_3 as per their sequential position. In other words, the numbers in the center for each of the three crosses will be $c_1 + 1$, $c_2 + 2$, and $c_3 + 3$ respectively as described below:

    The first cross:

    Add the four surrounding numbers: $c_1 = 5 + 2 + 1 + 3 = 11$

    Add 1: $c_1 + 1 = 11 + 1 = 12$. This equals the number in the center.

The second cross:

Add the four surrounding numbers: $c_2 = 1 + 2 + 4 + 4 = 11$

Add 2: $c_2 + 2 = 11 + 2 = 13$. This equals the number in the center.

The third cross:

Add the four surrounding numbers: $c_3 = 3 + 4 + 2 + 5 = 14$

Add 3: $c_3 + 3 = 14 + 3 = 17$. This is the answer we found assuming a sequential dependency among the crosses. As you will see later, this is not the right way to design such problems.

For each cross, we have added natural numbers (1, 2, and 3) to the sum of four surrounding numbers. In some other problems, an entirely different set of sequential numbers (e.g., 0, 1, 3, or -1, -2, -3, or 2, 3, 4, etc.) could be added to the values that result from applying different arithmetic operations on the numbers in the cross. This is what we mean by having a sequential dependency among the crosses.

While it is possible that some books or aptitude tests might require problems to be solved this way, we believe that all the logic to determine the number should be encoded in the individual cross, i.e., each cross should contain the logic within itself. Hence, the answers in this book do not assume any form of sequential dependency among the crosses. However, readers are advised to follow the *guidelines and norms* of the books being read or the tests being taken to solve the problems in those books or tests and prepare accordingly.

But, can the problem in figure-B1 be solved without using

sequential dependency at all? In cases when a problem is designed by assuming sequential dependency, finding the solution in other ways might not even be considered. However, in this instance we can find an answer without using any sequential dependency as explained below:

✔ Add the two numbers in the horizontal row and let us call the total A.

✔ Multiply the two numbers in the vertical column and let us call the result B.

✔ As shown below, the number in the center is A + B:

The first cross:

$A = 5 + 1 = 6$

$B = 2 \times 3 = 6$

$A + B = 6 + 6 = 12$. This equals the number in the center.

The second cross:

$A = 1 + 4 = 5$

$B = 2 \times 4 = 8$

$A + B = 5 + 8 = 13$. This equals the number in the center.

The third cross:

$A = 3 + 2 = 5$

$B = 4 \times 5 = 20$

$A + B = 5 + 20 = 25$. This is the answer to the problem.

This answer differs from what we found earlier by assuming the sequential dependency among the crosses. However, did you notice how we solved this problem without assuming any sequential dependency at all? The crosses were merely using the same logic to determine the missing number, which is the right way to design such type of problems. Therefore, we will not consider any sequential dependency among the crosses when solving problems in this book.

(C) Unknown Variables and Coefficients

Here we will discuss *Unknown Variables* and *Coefficients*. Let us take an example:

Example-1: The price of 3 pencils is 15, what is the price of one pencil?

Answer: We can write the problem as :

3 × price of one pencil = 15

Hence, the price of one pencil = $\dfrac{15}{3}$ = 5

In Mathematical Form: Let us write this in a mathematical form. In mathematics, we use the letter "x" to denote an unknown variable. Let us assume that the price of one pencil is x.

Then the price of 3 pencils = 3 × price of one pencil = $3x$.

But we know that the price of 3 pencils is 15, hence we can write the following expression:

$$3x = 15$$

$$x = \dfrac{15}{3} = 5$$

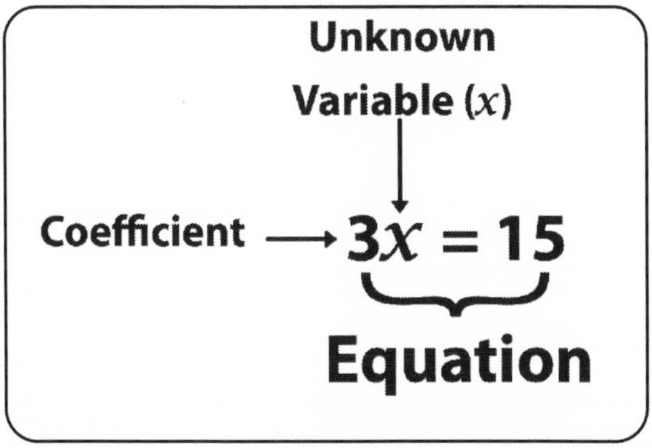

Figure-C1: Equation, Unknown variable, and Coefficient

In the above example:

- $3x = 15$ is an equation, whose right side is 15 and the left side is $3x$. Refer to **Appendix-D** for more details about *Equations*.

- A variable is a quantity that changes its value in the context of a problem. x is an unknown variable because we don't know the value of this variable. In this equation, we are solving for the unknown variable x.

- 3 is the coefficient of x.

This completes our brief introduction to the topic.

(D) Solving Linear Equations

Let us start with the definition of an *equation*.

♛ Definition - Equation

An *equation* contains numbers and variables and has two sides: LHS (*left-hand side*) and RHS (*right-hand side*). The two sides are separated by an equal ("=") sign. Here is an example of an equation:

$$2x + 5 = 11$$

In the equation above, $(2x + 5)$ is the left-hand side (LHS) and 11 is the right-hand side (RHS) of the equation. An equation is either true or false depending on the value of the variable. In the equations above, we will see that for some value of x, the equation is true, i.e., LHS = RHS. Such values of x are called the solutions of the equation.

$$x = 1: \text{RHS} = 2x + 5 = 2 \times 1 + 5 = 7 \neq \text{RHS}$$

$$x = 2: \text{RHS} = 2x + 5 = 2 \times 2 + 5 = 9 \neq \text{RHS}$$

$$x = 3: \text{RHS} = 2x + 5 = 2 \times 3 + 5 = 11 = \text{RHS}$$

Therefore, $x = 3$ is the solution of the equation, whereas 1 and 2 are not.

♜ Note: Number of solutions

An equation can have *more than one* solution. For example, a quadratic equation[1] can have two solutions. In some cases, an equation can have an infinite number of solutions or have no

---

1. Refer to any standard textbook on *Quadratic Equation* for further details.

solutions at all.

For example :

(i) Equation $x = x$ is true for all values of x, hence it has an infinite number of solutions.

(ii) Equation $x^2 = x$ is true for two values of x (0,1), hence it has two solutions.

(iii) Equation $x + 6 = 3x$ is true for one values of $x = 3$, hence it has one solution.

(iv) Equation $x + 2 = x + 5$ is not true for any value of x, hence it has no solution.

♛ Definition: Linear Equation

If the highest power of a variable in the equation is 1, it is called a *linear equation.* Here are a few examples of linear equations:

(i) $2x + 3y = 3$

(ii) $2x + 5 = 8$

(iii) $2x + 3y + 5z = 9$

However, $2x^2 + 5x - 3 = 0$ is not a linear equation, as the highest power of the variable x is 2, and so it is a quadratic equation.

In the section below, we will discuss the two methods of solving *linear equations* involving two variables:

(A) Substitution

In this method, we substitute the variable with another variable for finding the solution. Let us take an example of two linear equations with two variables:

$$x + 3y = 9 \qquad \text{Equation (1)}$$

$$2x + 5y = 17 \qquad \text{Equation (2)}$$

Here is a step-by-step approach to solving this problem:

✔ **Step-1**: Now we choose one of the equations, and solve the equation for either variable, as shown below:

$$x + 3y = 9 \qquad \text{[Equation (1)]}$$

$$x = 9 - 3y \qquad \text{Equation (1a)}$$

✔ **Step-2**: We substitute the value of x in equation (2) to determine the value of y:

$$2x + 5y = 17 \qquad \text{[Equation (2)]}$$

Using $x = 9 - 3y$ from equation(1a)

$$2(9 - 3y) + 5y = 17$$

$$18 - 6y + 5y = 17$$

$$18 - y = 17$$

$$y = 18 - 17$$

$$y = 1$$

✔ **Step-3**: We now use the value of y in equation (1a) to find the value of x as shown below:

$$x = 9 - 3y \qquad \text{[Equation (1a)]}$$

$$= 9 - 3 \times 1 \qquad \text{[Using } y = 1\text{]}$$

$$= 9 - 3 = 6$$

Hence, $x = 6$

So, the solution of the equations (1) and (2) is $x = 6$, $y = 1$.

(B) Elimination

In this method, we transform one or both equations such that one of the variables cancels out by adding or subtracting the transformed equations. Let us take equations (1) and (2) again to solve using the *elimination method*.

$$x + 3y = 9 \qquad \text{Equation (1)}$$

$$2x + 5y = 17 \qquad \text{Equation (2)}$$

✔ **Step-1**: If we multiply equation (1) by 2, we get a transformed equation as shown below:

$$x + 3y = 9 \qquad \text{[Equation (1)]}$$

Multiply both sides by 2:

$$2(x + 3y) = 2 \times 9$$

$$2x + 6y = 18 \qquad \text{[Equation (1a)]}$$

✔ **Step-2**: Now we subtract equation (2) from equation (1a):

$$2x + 5y = 17 \qquad \text{Equation (2)}$$

$$2x + 6y = 18 \qquad \text{[Equation (1a)]}$$

Subtracting both sides, the terms containing x cancel out each other:

$$(2x + 6y) - (2x + 5y) = 18 - 17$$

$$2x + 6y - 2x - 5y = 1$$

$$y = 1$$

We can now use the value of y in equation (1) or (2) to find the value of x. Let us choose equation (2):

$$2x + 5y = 17 \qquad \text{[Equation (2)]}$$

$$2x + 5 \times 1 = 17 \qquad \text{[Using value of } y = 1\text{]}$$

$$2x = 17 - 5 = 12$$

$$x = \frac{12}{2}$$

$$x = 6$$

So, the solution of the equations (1) and (2) is $x = 6$, $y = 1$, which is the same as what we found using the *substitution method*. In this book, whenever appropriate, we will prefer one method over the other.

Independent equation

Let us discuss another concept called "**Independent equation**." Here is the rule:

> *If we have N unknown variables, we need N independent equations to find a single unique solution.*

For example, we need 2 independent equations to solve a system of equations containing 2 variables. In a system of simultaneous equations, an *independent equation* is the one which cannot be algebraically derived from other equations.

Example-1: The following two equations are not independent, because if we multiply equation (i) by 2, we will get equation (ii):

(i) $2x + 3y = 6$

(ii) $4x + 6y = 12$

Example-2: The following equations are not independent, because if we add equation (i) and equation (ii), we will get equation (iii):

(i) $x + 2y + 3z = 6$

(ii) $3x + 7y + z = 9$

(iii) $4x + 9y + 4z = 15$

Example-3: Following two equations are independent, as we cannot derive (i) from (ii) and vice-versa.

(i) $x + 3y = 9$

(ii) $2x + 5y = 17$

We have already found the solution for these equations as shown below:

Solution: $x = 6$ and $y = 1$

♖ **Note: Solving equations involving more than two variables**

To solve linear equations with more variables, we first write a system of linear equations and then solve them using *Cramer's rule*. Please refer to any standard algebra textbook for further details on this.

As the number of variables in a set of equations increases, finding the solutions manually becomes lengthy and cumbersome and is prone to errors. Hence, it is generally advisable to use an appropriate computer program or software (e.g., python, excel, R, or Matlab, etc.) to solve equations involving more than three variables. This completes our brief introduction to *linear equations*. For further details, refer to any standard textbook on this topic.

(E) Pascal's Triangle and Number Pyramid

In this section, we are going to derive the equations that will be used in solving *Number Pyramid problems*.

i. Equation for a 3-row pyramid

We can write the values of x_1, x_2 and x_2 as the sum of variables in the cells from the row directly below it:

$$x_1 = x_2 + x_3$$

$$x_2 = x_4 + x_5$$

$$x_3 = x_5 + x_6$$

3-Row pyramid

Using the value of x_2 and x_3 in the following equation:

$$x_1 = x_2 + x_3$$

$$= (x_4 + x_5) + (x_5 + x_6)$$

$$= x_4 + 2x_5 + x_6$$

Hence, the equation relating x_1 and x_4, x_5, x_6 is:

$$x_1 = x_4 + 2x_5 + x_6 \qquad \text{Equation (1)}$$

ii. Equation for a 4-row pyramid

We have already derived the following equation relating x_1 and x_4, x_5, x_6:

$$x_1 = x_4 + 2x_5 + x_6 \qquad \text{[Using equation (1)]}$$

We can write the values of x_4, x_5, and x_6 as the sum of variables in the cells from the row directly below it:

$$x_4 = x_7 + x_8$$

$$x_5 = x_8 + x_9$$

$$x_6 = x_9 + x_{10}$$

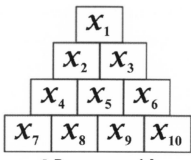

4-Row pyramid

Using the above in equation (1):

$$x_1 = x_4 + 2x_5 + x_6$$

$$= (x_7 + x_8) + 2(x_8 + x_9) + (x_9 + x_{10})$$

$$= x_7 + x_8 + 2x_8 + 2x_9 + x_9 + x_{10}$$

$$= x_7 + 3x_8 + 3x_9 + x_{10}$$

Hence, the equation relating x_1 and x_7, x_8, x_9, x_{10} is:

$$\mathbf{x_1 = x_7 + 3x_8 + 3x_9 + x_{10}}\quad \text{Equation (2)}$$

iii. Equation for a 5-row pyramid

We have already derived the following Equation relating x_1 and x_7, x_8, x_9, x_{10}:

$$x_1 = x_7 + 3x_8 + 3x_9 + x_{10}\quad \text{[Equation (2)]}$$

We can write the values of x_7, x_8, x_9 and x_{10} as the sum of variables in the cells from the row directly below it:

$$x_7 = x_{11} + x_{12}$$

$$x_8 = x_{12} + x_{13}$$

$$x_9 = x_{13} + x_{14}$$

5-Row pyramid

172

$$x_{10} = x_{14} + x_{15}$$

Using the above in equation (2):

$$x_1 = x_7 + 3x_8 + 3x_9 + x_{10} \qquad \text{[Equation (2)}$$

$$= (x_{11} + x_{12}) + 3(x_{12} + x_{13}) + 3(x_{13} + x_{14}) + (x_{14} + x_{15})$$

$$= x_{11} + x_{12} + 3x_{12} + 3x_{13} + 3x_{13} + 3x_{14} + x_{14} + x_{15}$$

$$= x_{11} + 4x_{12} + 6x_{13} + 4x_{14} + x_{15}$$

Hence, the equation relating x_1 and $x_{11}, x_{12}, x_{13}\ x_{14}\ x_{15}$ is:

$$\mathbf{x_1 = x_{11} + 4x_{12} + 6x_{13} + 4x_{14} + x_{15}} \quad \text{Equation (3)}$$

The equations we derived above link the numbers from two different rows in a pyramid. Let us review this with an example.

Example: Find the values of A, B, and C in the pyramid.

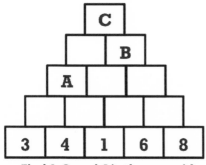

Find A, B, and C in the pyramid

Figure-E1: Problem

One way to solve this is to add all the numbers iteratively upward starting from the bottom row. Instead, we will use the equations derived earlier as shown below:

✔ To determine A, we will take 3, 4, and 1 from the bottom row and use the following equation :

$$x_4 = x_{11} + 2x_{12} + x_{13}$$

Using $x_4 = A, x_{11} = 3, x_{12} = 4, x_{13} = 1$

$$A = 3 + 2 \times 4 + 1 = 3 + 8 + 1 = 12$$

• To determine B, we will take 4, 1, 6, and 8 from the bottom row and use the following equation :

$$x_3 = x_{12} + 3x_{13} + 3x_{14} + x_{15}$$

Here $x_3 = B, x_{12} = 4, x_{13} = 1, x_{14} = 6, x_{15} = 8$

$$B = 4 + 3 \times 1 + 3 \times 6 + 8 = 4 + 3 + 18 + 8 = 33$$

• To determine C, we will take 3, 4, 1, 6, and 8 from the bottom row and use the following equation :

$$x_1 = x_{11} + 4x_{12} + 6x_{13} + 4x_{14} + x_{15}$$

Here $x_1 = C, x_{11} = 3, x_{12} = 4, x_{13} = 1, x_{14} = 6, x_{15} = 8$

$$C = 3 + 4 \times 4 + 6 \times 1 + 4 \times 6 + 8$$
$$= 3 + 16 + 6 + 24 + 8$$
$$= 57$$

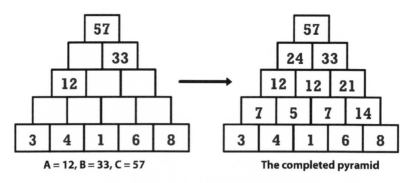

A = 12, B = 33, C = 57   The completed pyramid

Figure-E2: Answer

The values of A, B, C in the pyramid are shown in the figure on the right. Did you notice the benefit of using the equation to find the missing numbers in a pyramid? These equations can be used to find the desired number in a pyramid directly without having to add the numbers upwards. The same concept can also be used to formulate equations that can be used to solve the pyramid problems given in **Chapter-6 (p. 115)**.

It is also interesting to note that the coefficients in the equations derived earlier can also be derived from Pascal's triangle as illustrated in the figure below. Readers can refer to **Chapter-6 (p. 115)** for further details on this.

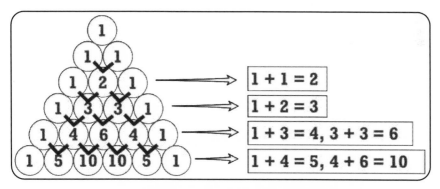

Figure-E3: Pascal's triangle

The concepts discussed in this section are useful for solving *Number pyramid*. Refer to *Number pyramid* in **Chapter-6 (p. 115)**. This completes a brief introduction to this topic.

Section-III Answers

This section contains answers to **Level-I** and **Level-II** questions provided in the exercise section of each chapter in the book.

Chapter 1 - Number Box

Level-I

(i)	-1	(vi)	11	(xi)	-2	(xvi)	6	(xxi)	-1
(ii)	45	(vii)	56	(xii)	150	(xvii)	36	(xxii)	9
(iii)	-14	(viii)	22	(xiii)	156	(xviii)	36	(xxiii)	-15
(iv)	2	(ix)	-8	(xiv)	19	(xix)	12	(xxiv)	90
(v)	28	(x)	97	(xv)	24	(xx)	8	(xxv)	15

Level-II

(i)	3	(vi)	127	(xi)	8	(xvi)	10	(xxi)	12
(ii)	41	(vii)	-0.5	(xii)	35	(xvii)	8	(xxii)	17
(iii)	41	(viii)	141	(xiii)	-0.25	(xviii)	37	(xxiii)	45
(iv)	40	(ix)	17	(xiv)	91	(xix)	20	(xxiv)	73
(v)	72	(x)	133	(xv)	6	(xx)	10	(xxv)	12

Chapter 2 - Number Cross

Level-I

(i)	17	(vi)	2	(xi)	10	(xvi)	23	(xxi)	4
(ii)	84	(vii)	8	(xii)	1	(xvii)	63	(xxii)	14
(iii)	20	(viii)	13	(xiii)	100	(xviii)	9	(xxiii)	6.5
(iv)	3	(ix)	0	(xiv)	16	(xix)	53	(xxiv)	44
(v)	21	(x)	12	(xv)	18	(xx)	4	(xxv)	17

Level-II

(i)	49	(vi)	126	(xi)	-3	(xvi)	3	(xxi)	7
(ii)	1	(vii)	12	(xii)	89	(xvii)	38	(xxii)	-16
(iii)	0	(viii)	32	(xiii)	33	(xviii)	15	(xxiii)	24
(iv)	2	(ix)	-1	(xiv)	-84	(xix)	-4	(xxiv)	-17
(v)	20	(x)	-21	(xv)	-4	(xx)	-28	(xxv)	50

Chapter 3 - Marbles in a Box

Level-I

#	◯	●	#	◯	●
(i)	4	8	(vi)	-5	21
(ii)	-19	8	(vii)	25	5
(iii)	-8	8	(viii)	3	12
(iv)	45	-15	(ix)	3	7
(v)	5	7	(x)	7	2

Level-II

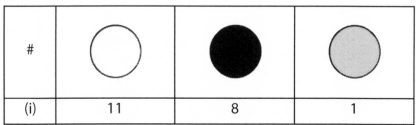

#	◯	●	◉
(i)	11	8	1

(ii)	36	-12	-11
(iii)	-17	16	-2
(iv)	9	2	-8
(v)	-40	20	30
(vi)	3	8	1
(vii)	11	14	-16
(viii)	9	3	1
(ix)	3	3	4
(x)	-26	12	6

Chapter 4 - Average cell

Level-I

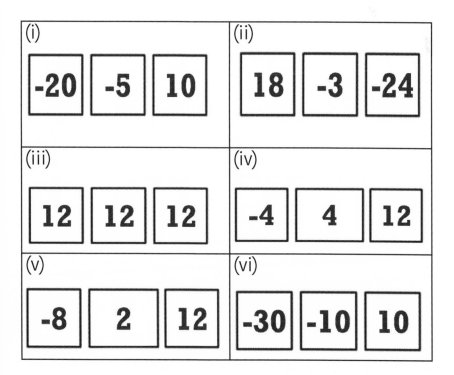

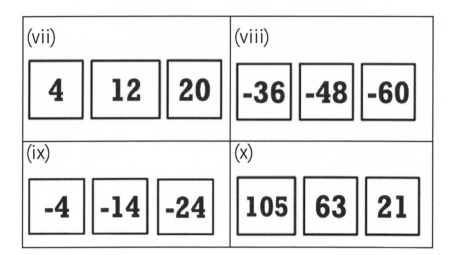

(vii)

| 4 | 12 | 20 |

(viii)

| -36 | -48 | -60 |

(ix)

| -4 | -14 | -24 |

(x)

| 105 | 63 | 21 |

Level-II

(i)

| 7 | 7 |
| 7 | 7 |

(ii)

| 6 | 6 | 8 |
| 4 | | |

(iii)

| 3 | 4 |
| 5 | 6 |

(iv)

| 4 | 8 | 4 |
| 16 | | |

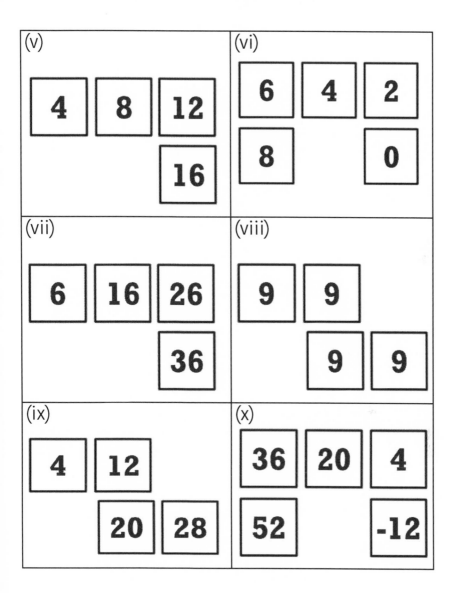

(v)

4	8	12
		16

(vi)

6	4	2
8		0

(vii)

6	16	26
		36

(viii)

9	9	
	9	9

(ix)

4	12	
	20	28

(x)

36	20	4
52		-12

Chapter 5 - Wisgo Number Tile

Level-I

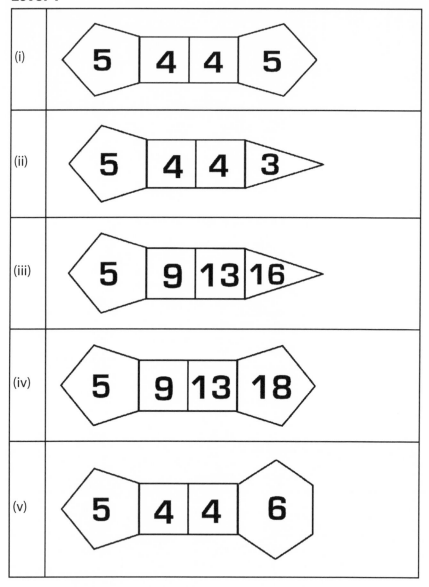

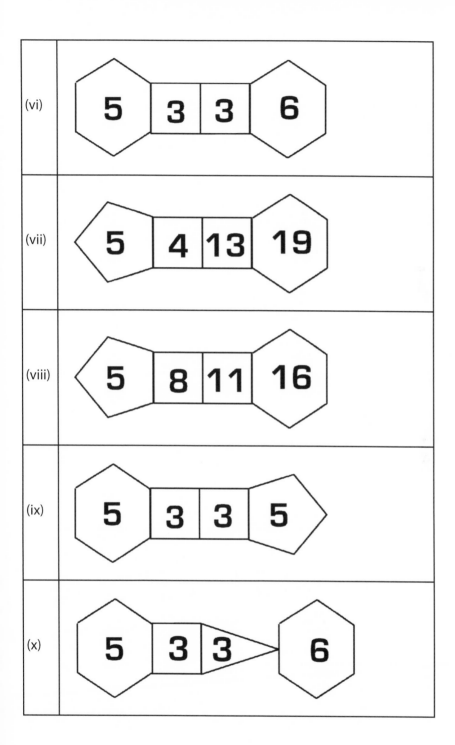

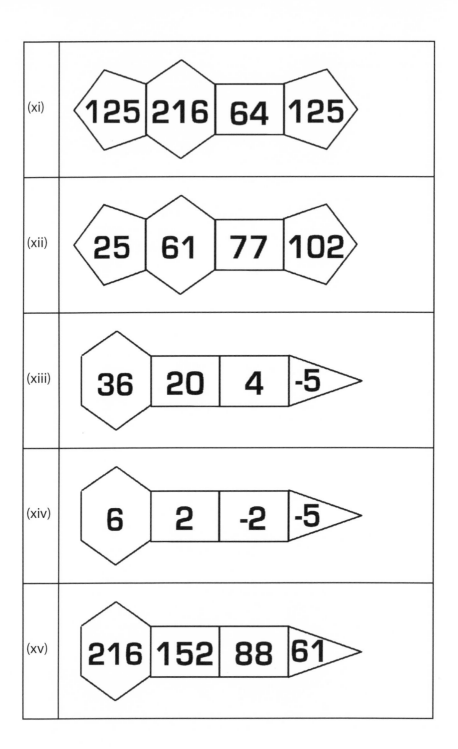

(xi) 125 216 64 125

(xii) 25 61 77 102

(xiii) 36 20 4 -5

(xiv) 6 2 -2 -5

(xv) 216 152 88 61

Level-II

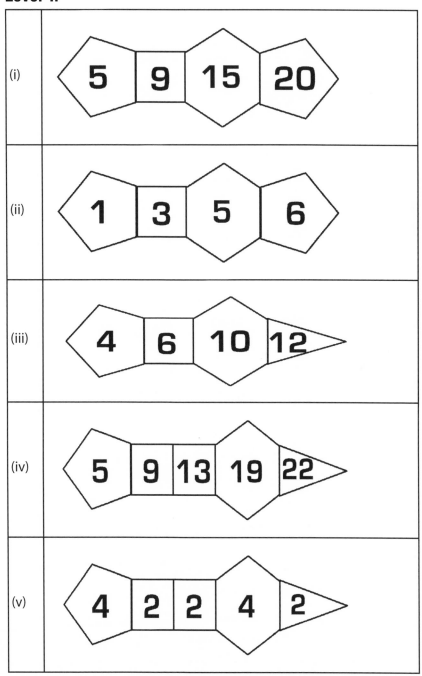

(i)	5 9 15 20
(ii)	1 3 5 6
(iii)	4 6 10 12
(iv)	5 9 13 19 22
(v)	4 2 2 4 2

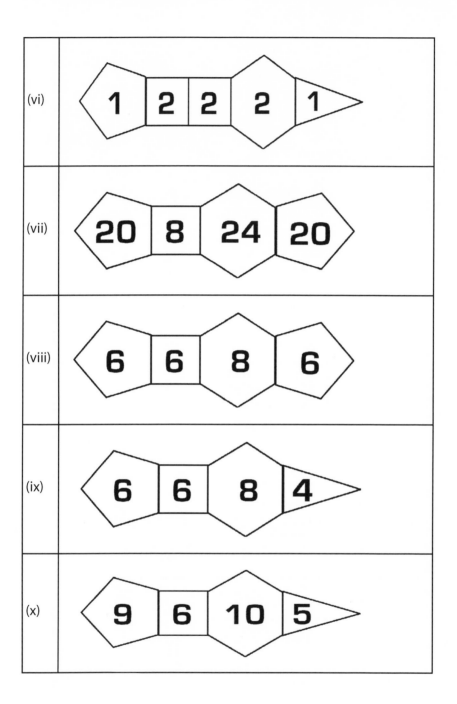

(vi) 1 2 2 2 1

(vii) 20 8 24 20

(viii) 6 6 8 6

(ix) 6 6 8 4

(x) 9 6 10 5

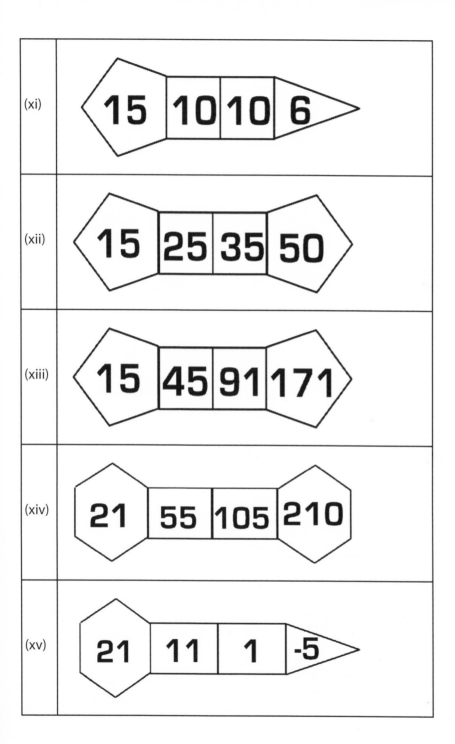

(xi) 15 10 10 6

(xii) 15 25 35 50

(xiii) 15 45 91 171

(xiv) 21 55 105 210

(xv) 21 11 1 -5

Chapter 6 - Number pyramid

Level-I

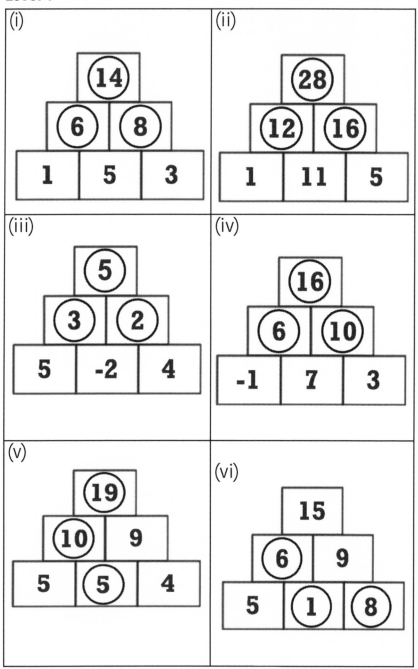

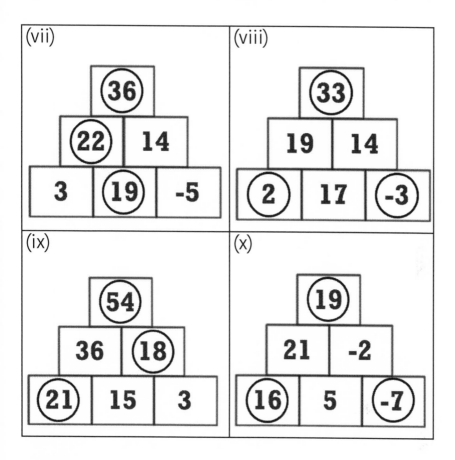

(vii)

```
        (36)
     (22)   14
   3   (19)   -5
```

(viii)

```
        (33)
     19    14
   (2)   17   (-3)
```

(ix)

```
        (54)
     36    (18)
   (21)   15    3
```

(x)

```
        (19)
     21    -2
   (16)   5   (-7)
```

Level-II

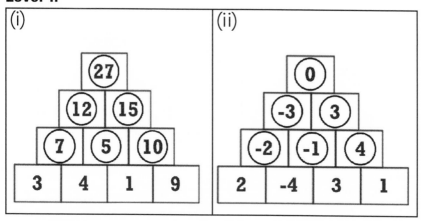

(i)

```
          (27)
       (12)  (15)
     (7)  (5)  (10)
   3    4    1    9
```

(ii)

```
          (0)
       (-3)  (3)
     (-2)  (-1)  (4)
   2    -4    3    1
```

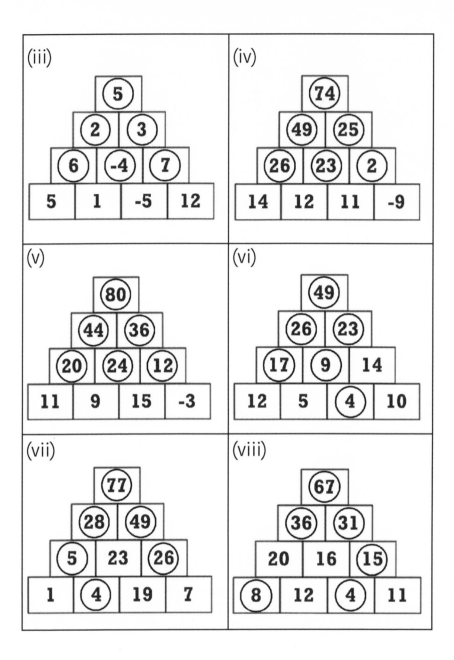

(iii)

```
            5
         2     3
      6    -4     7
    5    1    -5    12
```

(iv)

```
           74
        49     25
      26    23     2
    14    12    11    -9
```

(v)

```
           80
        44     36
      20    24    12
    11    9    15    -3
```

(vi)

```
           49
        26     23
      17     9    14
    12    5    4    10
```

(vii)

```
           77
        28     49
       5    23    26
    1    4    19    7
```

(viii)

```
           67
        36     31
      20    16    15
    8    12    4    11
```

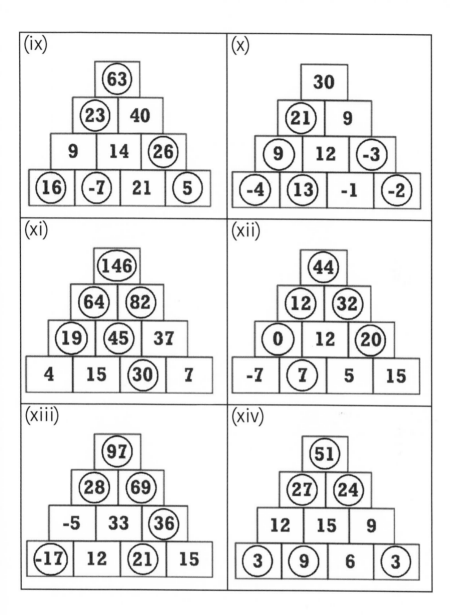

(ix)

```
            63
        23      40
      9    14    26
   16   -7   21    5
```

(x)

```
            30
        21      9
      9    12    -3
   -4   13   -1    -2
```

(xi)

```
            146
        64      82
      19   45    37
    4    15   30    7
```

(xii)

```
            44
        12      32
      0    12    20
   -7   7    5    15
```

(xiii)

```
            97
        28      69
      -5   33    36
   -17   12   21    15
```

(xiv)

```
            51
        27      24
      12   15    9
    3    9    6    3
```

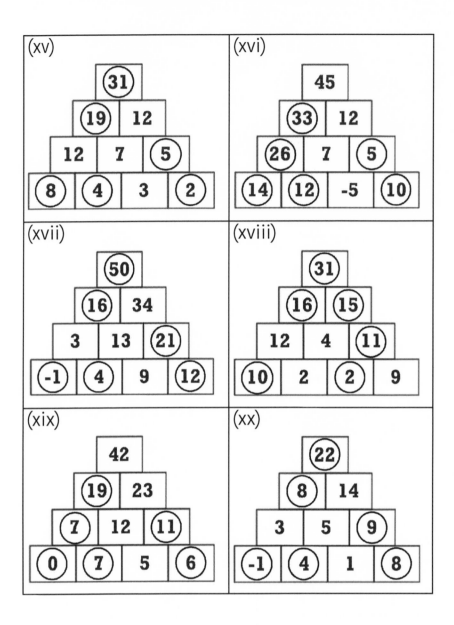

(xv)

```
          (31)
       (19)   12
     12    7   (5)
   (8)  (4)  3   (2)
```

(xvi)

```
          45
       (33)   12
     (26)   7   (5)
   (14) (12)  -5  (10)
```

(xvii)

```
          (50)
       (16)   34
      3    13   (21)
   (-1)  (4)  9   (12)
```

(xviii)

```
          (31)
       (16)  (15)
      12    4   (11)
   (10)  2   (2)  9
```

(xix)

```
          42
       (19)   23
     (7)  12  (11)
   (0)  (7)  5   (6)
```

(xx)

```
          (22)
       (8)   14
      3    5   (9)
   (-1)  (4)  1   (8)
```

Level-I

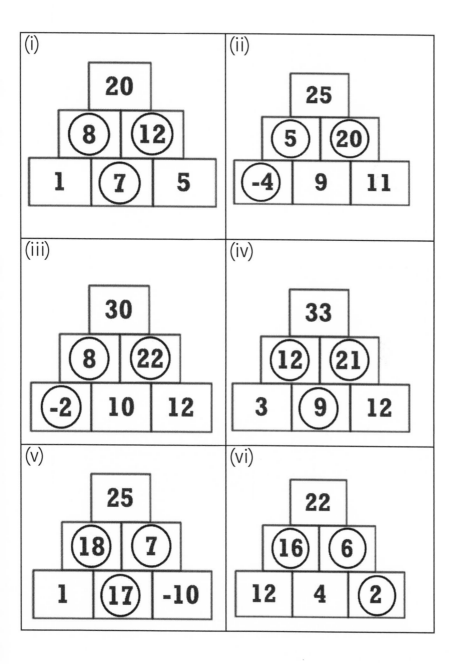

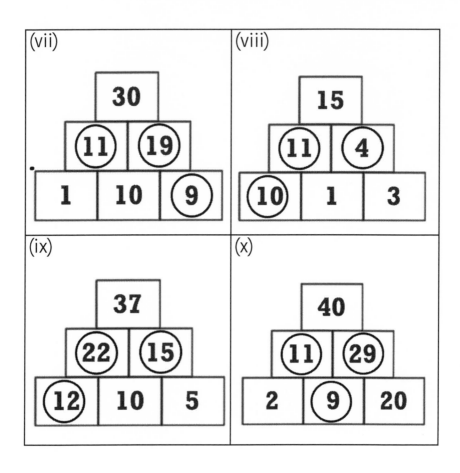

(vii)

30

(11) (19)

1 | 10 | (9)

(viii)

15

(11) (4)

(10) | 1 | 3

(ix)

37

(22) (15)

(12) | 10 | 5

(x)

40

(11) (29)

2 | (9) | 20

Level-II

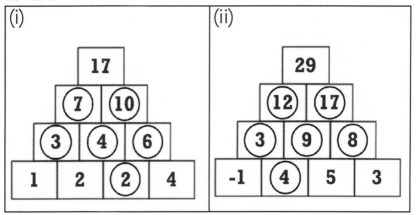

(i)

17

(7) (10)

(3) (4) (6)

1 | 2 | (2) | 4

(ii)

29

(12) (17)

(3) (9) (8)

-1 | (4) | 5 | 3

(iii)

47
(20) (27)
5 (15) 12
(-8) (13) 28 (-16)

(iv)

51
(19) (32)
(3) (16) (16)
1 2 (14) 2

(v)

(37)
15 (22)
(3) (12) 10
1 2 (10) (0)

(vi)

94
(57) (37)
(30) (27) 10
5 (25) 2 (8)

(vii)

85
(33) (52)
4 (29) 23
(-13) (17) 12 (11)

(viii)

(47)
19 (28)
(3) (16) 12
1 2 (14) (-2)

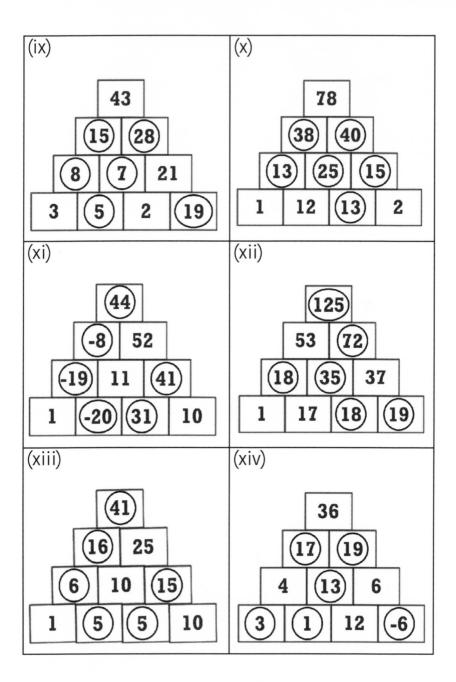

(ix)

43
(15) (28)
(8) (7) 21
3 (5) 2 (19)

(x)

78
(38) (40)
(13) (25) (15)
1 12 (13) 2

(xi)

(44)
(-8) 52
(-19) 11 (41)
1 (-20) (31) 10

(xii)

(125)
53 (72)
(18) (35) 37
1 17 (18) (19)

(xiii)

(41)
(16) 25
(6) 10 (15)
1 (5) (5) 10

(xiv)

36
(17) (19)
4 (13) 6
(3) (1) 12 (-6)

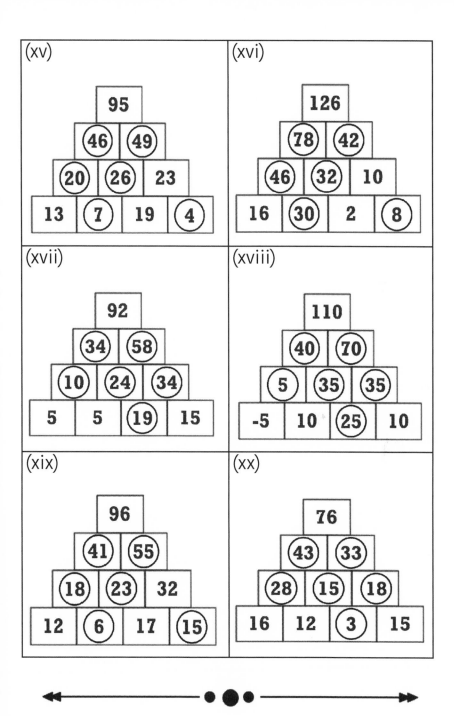

(xv)

```
            95
        46      49
      20    26    23
    13    7    19    4
```

(xvi)

```
            126
        78      42
      46    32    10
    16    30    2    8
```

(xvii)

```
            92
        34      58
      10    24    34
    5    5    19    15
```

(xviii)

```
            110
        40      70
      5    35    35
    -5    10    25    10
```

(xix)

```
            96
        41      55
      18    23    32
    12    6    17    15
```

(xx)

```
            76
        43      33
      28    15    18
    16    12    3    15
```

Made in the USA
Monee, IL
25 April 2020